237

Topics in Current Chemistry

Topics in Current Chemistry

Recently Published and Forthcoming Volumes

Natural Product Synthesis II
Volume Editor: Mulzer, J.H.
Vol. 244, 2004

Natural Product Synthesis I
Volume Editor: Mulzer, J.H.
Vol. 243, 2004

Immobilized Catalysts
Volume Editor: Kirschning, A.
Vol. 242, 2004

Transition Metal and Rare Earth Compounds III
Volume Editor: Yersin, H.
Vol. 241, 2004

The Chemistry of Pheromones and Other Semiochemicals II
Volume Editor:Schulz, S.
Vol. 240, 2004

The Chemistry of Pheromones and Other Semiochemicals I
Volume Editor:Schulz, S.
Vol. 239, 2004

Orotidine Monophosphate Decarboxylase
Volume Editors: Lee, J.K., Tantillo, D.J.
Vol. 238, 2004

Long-Range Charge Transfer in DNA II
Volume Editor: Schuster, G.B.
Vol. 237, 2004

Long-Range Charge Transfer in DNA I
Volume Editor: Schuster, G.B.
Vol. 236, 2004

Spin Crossover in Transition Metal Compounds III
Volume Editors: Gütlich, P., Goodwin, H.A.
Vol. 235, 2004

Spin Crossover in Transition Metal Compounds II
Volume Editors: Gütlich, P., Goodwin, H.A.
Vol. 234, 2004

Spin Crossover in Transition Metal Compounds I
Volume Editors: Gütlich, P., Goodwin, H.A.
Vol. 233, 2004

New Aspects in Phosphorus Chemistry IV
Volume Editor: Majoral, J.-P.
Vol. 232, 2004

Elemental Sulfur and Sulfur-Rich Compounds II
Volume Editor: Steudel, R.
Vol. 231, 2003

Elemental Sulfur and Sulfur-Rich Compounds I
Volume Editor: Steudel, R.
Vol. 230, 2003

New Aspects in Phosphorus Chemistry III
Volume Editor: Majoral, J.-P.
Vol. 229, 2003

Dendrimers V
Volume Editors: Schalley, C.A., Vögtle, F.
Vol. 228, 2003

Colloid Chemistry II
Volume Editor: Antonietti, M
Vol. 227, 2003

Colloid Chemistry I
Volume Editor: Antonietti, M
Vol. 226, 2003

Modern Mass Spectrometry
Volume Editor: Schalley, C.A.
Vol. 225, 2003

Hypervalent Iodine Chemistry
Volume Editor: Wirth, T
Vol. 224, 2003

Long-Range Charge Transfer in DNA II

Volume Editor: G. B. Schuster

With contributions by
D. Beratan · Y.A. Berlin · A.L. Burin · Z. Cai · E. Conwell
G. Cuniberti · R. di Felice · N.E. Geacintov · I.V. Kurnikov
D. Porath · M.A. Ratner · N. Rösch · M.D. Sevilla
V. Shafirovich · H.H. Thorp · A.A. Voityuk

Springer

The series *Topics in Current Chemistry* presents critical reviews of the present and future trends in modern chemical research. The scope of coverage includes all areas of chemical science including the interfaces with related disciplines such as biology, medicine and materials science. The goal of eaxch thematic volume is to give the nonspecialist reader, whether at the university or in industry, a comprehensive overview of an area where new insights are emerging that are of interest to a larger scientific audience.

As a rule, contributions are specially commissioned. The editors and publishers will, however, always be pleased to receive suggestions and supplementary information. Papers are accepted for *Topics in Current Chemistry* in English.

In references *Topics in Current Chemistry* is abbreviated Top Curr Chem and is cited as a journal.

Visit the TCC home page at http://www.springerlink.com/

ISSN 0340-1022
ISBN 978-3-662-14437-4 ISBN 978-3-540-39912-4 (eBook)
DOI 10.1007/978-3-540-39912-4

Library of Congress Catalog Card Number 74-644622

springeronline.com

Originally published by Springer-Verlag Berlin Heidelberg New York in 2004
Softcover reprint of the hardcover 1st edition 2004

Cover design: KünkelLopka, Heidelberg/design & production GmbH, Heidelberg
Typesetting: Stürtz AG, Würzburg

02/3020 ra – 5 4 3 2 1 0 – Printed on acid-free paper

Volume Editor

Dr. Gary B. Schuster
Georgia Institute of Technology
School of Chemistry and Biochemistry
Atlanta GA 30332-0400
USA
E-mail: gary.schuster@cos.gatech.edu

Editorial Board

Topics in Current Chemistry is also Available Electronically

For all customers who have a subscription to Topics in Current Chemistry, we offer the electronic version via SpringerLink free of charge. Please contact your librarian who can receive a password for free access to the full articles by registering at:

http://www.springerlink.com

If you do not have a subscription, you can still view the tables of contents of the volumes and the abstract of each article by going to the SpringerLink Homepage, clicking on "Browse by Online Libraries", then "Chemical Sciences", and finally choose Topics in Current Chemistry.

You will find information about the
- Editorial Board
- Aims and Scope
- Instructions for Authors
- Sample Contribution

at http://www.springeronline.com using the search function.

Preface

The central role played by DNA in cellular life guarantees a place of importance for the study of its chemical and physical properties. It did not take long after Watson and Crick described the now iconic double helix structure for a question to arise about the ability of DNA to transport electrical charge. It seemed apparent to the trained eye of the chemist or physicist that the array of neatly stacked aromatic bases might facilitate the movement of an electron (or hole) along the length of the polymer. It is now more than 40 years since the first experimental results were reported, and that question has been answered with certainty.

As you will earn by reading these volumes, Long-Range Charge Transfer in DNA I and II, today no one disputes the fact that charge introduced at one location in DNA can migrate and cause a reaction at a remote location. In the most thouroughly studied example, it is clear that a radical cation injected at a terminus of the DNA polymer can cause a reaction at a $(GG)_n$ sequence located hundreds of Ångstroms away.

In the last decade, an intense and successful investigation of this phenomenon has focused on its mechanism. The experimental facts discovered and the debate of their interpretation form large portions of these volumes. The views expressed come both from experimentalists, who have devised clever tests of each new hypothesis, and from theorists, who have applied these findings and refined the powerful theories of electron transfer reactions. Indeed, from a purely scientific view, the cooperative marriage of theory and experiment in this pursuit is a powerful outcome likely to outlast the recent intense interest in this field.

Is the quest over? No, not nearly so. The general agreement that charge can migrate in DNA is merely the conclusion of the first chapter. This hard-won understanding raises many important new questions. Some pertain to oxidative damage of DNA and mutations in the genome. Others are related to the possible use of the charge transfer ability of DNA in the emerging field of molecular-scale electronic devices. Still others are focused on the application of this phenomenon to the development of clinical assays.

It is my hope that these volumes will serve as a springboard for the next phase of this investigation. The foundation knowledge of this field contained within these pages should serve as a defining point of reference for all who explore its boundaries. For this, I must thank all of my coauthors for their effort, insight and cooperation.

Atlanta, January 2004 Gary B. Schuster

Contents

Contents of Volume 236

Long-Range Charge Transfer in DNA I

Volume Editor: Gary B. Schuster

ISBN 3-540-20127-0

Top Curr Chem (2004) 237:1–36
DOI 10.1007/b94471

DNA Electron Transfer Processes: Some Theoretical Notions

Yuri A. Berlin[1] · Igor V. Kurnikov[1, 2] · David Beratan[2] · Mark A. Ratner[1] · Alexander L. Burin[1, 3]

[1] Department of Chemistry, Center for Nanofabrication and Molecular Self-Assembly and Materials Research Center, Northwestern University, 2145 Sheridan Road, Evanston, IL 60208, USA
E-mail: aburin@tulane.edu
[2] Department of Chemistry, Duke University, Durham, NC 27708, USA
[3] Department of Chemistry, Tulane University, New Orleans, LA 70118, USA

Abstract Charge motion within DNA stacks, probed by measurements of electric conductivity and by time-resolved and steady-state damage yield measurements, is determined by a complex mixture of electronic effects, coupling to quantum and classical degrees of freedom of the atomic motions in the bath, and the effects of static and dynamic disorder. The resulting phenomena are complex, and probably cannot be understood using a single integrated modeling viewpoint. We discuss aspects of the electronic structure and overlap among base pairs, the viability of simple electronic structure models including tight-binding band pictures, and the Condon approximation for electronic mixing. We also discuss the general effects of disorder and environmental coupling, resulting in motion that can span from the coherent regime through superexchange-type hopping to diffusion and gated transport. Comparison with experiment can be used to develop an effective phenomenological multiple-site hopping/superexchange model, but the microscopic understanding of the actual behaviors is not yet complete.

Keywords Electron transfer · Hole transport · Hopping · Superexchange · Coupling to the molecular surroundings

Abbreviations and Symbols

a	Spacing between repeating units of the bridge
A	Adenine
A	Polarization matrix
A_{ij}	Elements of the polarization matrix
b	Transfer integral
B	Bridge connecting a donor and an acceptor
β	Falloff parameter for the distance dependence of the electron transfer rate
c_i	Annihilation operator for a hole at the *i*-th site of the chain describing the stack of Watson–Crick base pairs
c_i^+	Creation operator for a hole at the *i*-th site of the chain describing the stack of Watson–Crick base pairs
C	Cytosine
D	Width of the rectangular barrier
D–A	Donor–acceptor tunneling
(DWFC)	Density of states weighted Franck–Condon factor
DNA	Deoxyribonucleic acid
Δ_b	Barrier height for the adiabatic hole motion
ΔE	Difference in ionization potentials of adenine–thymine and guanine–cytosine base pairs
ΔE_b	Energy barrier between the injection energy and the barrier height
ΔG^0	Driving force for electron transfer
ET	Electron transfer

E	Energy of the particle undergoing a tunneling transition through the rectangular barrier
E_{B_i}	Electronic energy of the bridge state $\|B_i\rangle$
E_{tun}	Electronic energy associated with the "transfer electron" in the activated complex
E_v	Energy of the ν-th vibrational state
ε	Dielectric constant of the solvent
g	Conductance
G	Guanine
$\hbar$	Planck constant
H_{DA}	Effective donor–acceptor interaction
HOMO	Highest occupied molecular orbital
k	Rate constant of electron transfer
k_B	Boltzmann constant
k_0	Pre-exponential factor in Eq. 6 for the rate of the elementary hopping step
L	Length of the bridge containing adenine–cytosine base pairs only
LUMO	Lowest unoccupied molecular orbital
λ	Marcus reorganization energy
m	Mass of the tunneling particle
n_i	Population of i-th site of the chain describing the stack of Watson–Crick base pairs
N	Number of sites through which the electron or hole tunnels
NDO SCF	Neglect of differential overlap self-consistent field method
P_G	Products formed in the reactions of water with guanine radical cation G_j^+
P_{GGG}	Product formed in the reaction of water with the hole trapped by the guanine triple GGG
P_v	Probability of the system to be found in the vibrational state v
ω_v	Effective vibronic frequency of the medium
q	Number of base pairs in the adenine–thymine bridge between two guanine sites
r	Spatial donor–acceptor separation
r_0	Spatial donor–acceptor separation in the certain reference state
$\rho_{FCv}(E)$	Generalized Franck–Condon factor
S_{vw}	Franck–Condon overlap factor
σ_0	Conductivity prefactor
T	Thymine
T	Temperature
τ_{LB}	Landauer–Buttiker tunneling time for the rectangular barrier
$\tau_{LB\text{-}M}$	Landauer–Buttiker tunneling time in a molecular orbital representation
τ_t	Tunneling time
U	Height of the rectangular barrier through which the particle is tunneling

$<V^2>$	Average squared electronic mixing between donor and acceptor
V_{BiA}	Hamiltonian term describing the interaction between the bridge state $\|B_i\rangle$ and the acceptor state $\|A\rangle$
V_{DBi}	Hamiltonian term describing the interaction between the donor state $\|D\rangle$ and the bridge state $\|B_i\rangle$
V_{rp}	Half of the effective energy splitting for the electron transfer reaction
v	Set of vibronic states that modulates the electron coupling matrix element
w	Set of vibronic states that does not modulate the electron coupling matrix element
x	Average position of a hole on the chain describing the stack of Watson–Crick base pairs
X_k	Multidimensional coordinate characterizing the polarization of water molecules
X_{opt}	Optimal value of the multidimensional coordinate characterizing the polarization of water molecules

1 Introduction

Electron transfer reactions are among the most widespread and significant in all of chemistry. Electron transfer (ET) within the double helical structure of DNA exhibits an extremely broad range of mechanistic behavior, and its exploration has become a focal point within the chemical community since the key studies of Barton and collaborators [1–3].

The current chapter discusses mechanisms of charge transfer reactions in double-stranded DNA (we will not deal with single-stranded DNA, or with individual bases or base-pair structures). We will also focus on interpretation of excess charge behavior in DNA molecules in terms of accepted theoretical models.

Figure 1 presents a very schematic picture of the DNA double helix, from the viewpoints of physical structure and a model for electronic behavior. Each base pair represents a localization site (actually, the localization is not on a base pair but on an individual base; because of the standard GC/AT hybridization, the language of base-pair localization is often used, when single-base localization is meant). Based on a wealth of evidence, both experimental and theoretical, we have indicated in Fig. 1b that the GC base pair is a more probable place for the hole to be localized—that is, it is easier to oxidize a G than any of the other three DNA bases. In the picture of Fig. 1b, each site (base pair) is assigned a unique energy, although it is clearly true that the energies will be modified by the neighbors with which the individual base pair interacts.

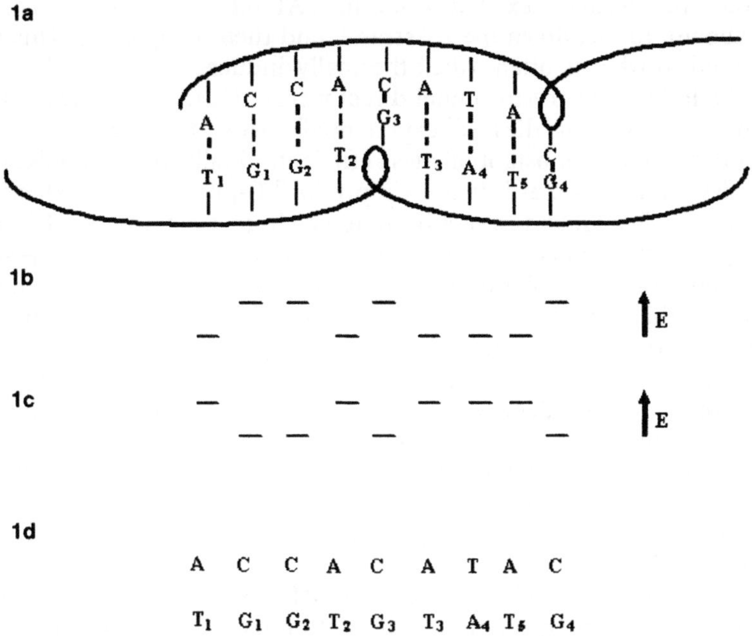

Fig. 1 Schematic illustration of the DNA double helical structure (**a**) and two possible mechanisms (electron-level energies in **b**, hole energies in **c**) of electronic motion in this molecule

Because most charge transfer processes and measurements in DNA actually consist of the motion of an electronic hole (that is, of a positive charge, corresponding to an ionized DNA base), it is more usual and more convenient to replace the picture in Fig. 1b by that of Fig. 1c. The diagram actually shows the energy of the holes, and indicates that the hole is more stable (lower lying energy level) on the GC pair than the AT pair. While this representation is confusing at first, it is both more common in the literature and more useful, and therefore we will depend upon it.

The simplest three mechanisms for motion, then, can be described in terms of the site picture in Fig. 1c. The sites G_1 and G_2 are located next to one another, and are separated by roughly 3.4 Å. Electrons can then tunnel between these two sites due to the overlap of the π-electron wave functions on the two (nearly cofacial) Gs. To move the electron from G_2 to G_3, it has to pass through T_2. Because T_2 is substantially higher in energy (the numbers differ, but a characteristic value between 0.2 and 0.4 eV is suspected), most of the hole wave functions will be localized on the Gs, with very little overlap onto the T site. Therefore, the motion from G_2 to G_3 is best represented as tunneling assisted by the presence of the T bridge, or as superexchange tunneling.

To move from G_3 to G_4, it is necessary for the hole to pass through three AT pairs. We can imagine this might happen in several different ways: the hole could be thermally excited to the first AT (energetically very costly, as noted above), tunnel down the AT strand, and then decay to G_4. This would correspond to what is often called thermally induced hopping [4]. Alternatively, the hole could try to tunnel directly from G_3 to G_4, but the extent of overlap charge mixing dies off exponentially with distance, and therefore this route should be substantially less efficient. Finally, the hole might actually be delocalized so that it is not simply "on" G_3, but actually extends over G_3, G_1, G_2, T_2, T_3, T_4, and even a bit onto G_4. In this picture, the delocalized hole migrates from having its center on G_3 to have its center on G_4—this is usually referred to as polaron hopping, although the term can be confusing (essentially, there can be small (localized) and large (delocalized) polarons, so that the term polaron motion needs to be more precisely qualified). The three mechanisms of tunneling, hopping, and thermally induced hopping differ in their distance dependence, their temperature dependence, and their rates. Direct tunneling falls off exponentially with distance as does superexchange tunneling; hopping is expected to fall off very slowly with length, as is thermally induced hopping. Tunneling and superexchange should depend on temperature much more weakly as compared with hopping and thermally induced hopping.

Because of the controlled disorder in DNA, all three of these processes can and do occur. Interpretation of any set of experiments, therefore, might be attempted using any of these mechanisms. Careful contrasting between experiment and model is quite necessary, to understand which of the possible charge transfer schemes in fact occurs for a given measurement on a given system under given conditions.

DNA electron transfer centers on two exponential factors: the Boltzmann population of electronic excited states and the distant-dependent probability of electron tunneling. The mechanism of electron transfer is determined by the relative value of these terms. The tunneling factor (for single-step D–A tunneling through the bridge) drops exponentially with the distance. This distance dependence is usually discussed in terms of the falloff parameter β. The Boltzmann factor (for carrier injection) drops exponentially with the energy mismatch between D–A and bridge states. The interplay of these rapidly varying factors defines the transport rate and mechanism, as well as their dependence on the DNA, donor, and acceptor.

2 Computation of Electronic Matrix Elements. Geometry and Energy Dependence

2.1 Preliminaries

Electron transfer through a molecular bridge can occur by single or multiple-step mechanisms [5–7]. The multiple steps may involve real (hopping) or virtual (superexchange) bridging states. Single electron transfer steps can occur in the strongly coupled (adiabatic) or weakly coupled (nonadiabatic) regimes [7]. In the weak-coupling regime, the electronic structure—including energetics and symmetry of the donor, acceptor, and bridge—determines the ET rate. In addition, the nonadiabatic rate depends on an activation free energy. In contrast, adiabatic ET is controlled by the activation free energies and dynamics of nuclear relaxation.

Transitions between these tunneling and hopping regimes have been probed in molecular wires, and DNA electron transfer systems generally reside near the boundary between these regimes. Whether DNA electron transfer is adiabatic or nonadiabatic depends upon the donor–acceptor distance. Since nearest-neighbor bases in DNA have electronic interaction energies as large as tenths of eV (see below), nearest-neighbor ET is likely adiabatic or nearly adiabatic.

Our attention here focuses on the change in donor–acceptor interaction strength as the number of bases intervening between donor and acceptor grows. In two-state donor–acceptor ET, the coupling matrix element is required for geometries at or near the activated complex in which the donor and acceptor localized electronic states are quasi-degenerate. There are several practical computational schemes to find this energy splitting [7–9]. While these specific schemes are described in greater detail below, it is instructive to introduce a perturbation theory that is commonly employed to lift degeneracy between two states when they do not interact directly with each other [10]. When a state $|D\rangle$ interacts indirectly with $|A\rangle$ through a manifold of $|B_i\rangle$ states, and the Hamiltonian term V describes the D–B and B–A interactions, the effective D–A interaction with a single superexchange step is [5–9]

$$H_{DA} = \sum_i V_{DB_i} V_{B_iA} / (E_{tun} - E_{B_i}) \tag{1}$$

Here E_{tun} is the electronic energy associated with the "transfer electron" in the activated complex. The orbital symmetry dependence arises from the V terms in the numerator. The energy dependence of the coupling—the energy mismatch between the D/A and bridge states—is reflected in the denominator. Note that when donor, acceptor, and bridge are in near resonance, the coupling changes rapidly with detuning of the bridge state energies from resonance with the tunneling energy. When the mismatch energy is large, there is only a weak dependence of coupling on tunneling energy.

2.2
Computing Coupling Elements

There are two qualitative ways to influence H_{DA}. One is to modify the V elements. The V elements are changed by altering the donor–bridge and acceptor–bridge interactions. Pulling D and A away from the helix, for example, weakens the V elements. In DNA, if π-orbital-mediated coupling dominated, moving the donor and acceptor from sites of intercalation to ribose-phosphate positions should weaken the V elements.

Simple changes in donor or acceptor distance to the DNA scale the V elements. Examples can be imagined associated with pulling a DNA groove-binding molecule away from the groove, or pulling an end-bound intercalator along the helical axis. These changes would scale down the value of H_{DA} without changing its dependence on distance. This difference between prefactor scaling and distance dependence is clearer when Eq. 1 is rewritten in the limit of weak interactions between bridging units, assuming only one orbital per bridge. This is the McConnell-like limit [7, 9]:

$$H_{DA} = V_{D1} V_{NA} / (E_{tun} - E_{B_1}) \prod_{i=1}^{N-1} V_{i,i+1} / (E_{tun} - E_{i+1}) \tag{2}$$

The tunneling energy in the denominator, E_{tun}, is the electronic energy associated with the nonequilibrium geometry donor and acceptor that are quasi-degenerate in energy in the "activated complex" of the ET system. In many electron transfer situations, including some DNA examples, the rate is found to decrease exponentially with distance:

$$k(r) = k(r_0) \exp[-\beta(r - r_0)] \tag{3}$$

with k, r, r_0, and β, respectively, the measured rate constant, the DA separation, the DA separation in some reference state, and the falloff parameter. The approximately exponential decay of the coupling with distance arises from the quotient terms in the product in Eq. 2. If all V elements are identical, $\beta = (-1/a) \ln |V_i/(E_{tun} - E_i)|$ where a is the spacing between repeating units of the bridge.

It is important to note that changes in the Vs need not cause a change in the falloff parameter β, but they may. For example, if the first and last steps on the tunneling pathways are simply weakened, but the π-stack dominates the propagation, the beta value will be unchanged by the modification. However, if V_{D1} and V_{NA} decrease sufficiently, the coupling through the π-stack will not influence the rate and the ribose phosphate or water-mediated coupling will dominate the decay. At very long distance, of course, the smallest β coupling pathway will dominate any tunneling (superexchange) mediation.

There are several methods in use to compute H_{DA} values. The most direct approach, applied widely to high-symmetry structures, is to compute a symmetric–antisymmetric splitting energy. However, in low-symmetry structures, this approach is not directly applicable; rather, the "minimum energy splitting" associated with the crossing between the potential energy surfaces is computed by driving the system through quasi-degeneracy (although this approach can be problematic) [8, 9]. Other approaches compress the many-orbital problem to an effective two-level system as suggested by Eq. 2 [5–9]. Recent studies of model π-stacks and DNA have compared the results arising from various calculation schemes [11, 12].

2.3 Ab Initio and Semiempirical Approaches to Donor–Acceptor Interactions in π-Stacks

The chemical control of DNA-mediated tunneling is apparent from Eq. 2: strong couplings among mediating groups and near resonance among donor, acceptor, and bridge states enhance H_{DA}. A proper description of coupling depends upon treating numerators and denominators in Eqs. 1 and 2 appropriately. The numerators depend on describing the wave function tails adequately, while the denominator depends on the quality of the energetics. Extended-Hückel, neglect of differential overlap (NDO) SCF, and ab initio Hartree–Fock methods have been used to compute interactions among DNA bases, to compute backbone vs base mediation effects, and to estimate the β values for DNA.

In the case of simple stacked aromatics, we find that NDO methods overestimate β values (by about 20%) compared to ab initio methods using standard split-valence basis sets, presumably because of the overly rapid decay of through-space propagation [11]. If π–π separation distances are greater than ~4 Å, the use of diffuse functions in the basis set is required. Both semiempirical and ab initio methods indicated that π-stack interactions dominate coupling for intercalated donors and acceptors. The case is more complex for backbone-attached donors and acceptors, where either the π-stack or the backbone may dominate, depending upon distance and attachment mode [12–15].

2.4 Electronic Coupling Through DNA

The first computations of DNA-mediated coupling matrix elements in modified DNAs were performed by Beratan and coworkers in 1993 [13] and computations were carried out on explicit experimental systems first in 1996 [14–15]. These results indicated a familiar U-shaped tunneling energy dependence of H_{DA} [16] (arising from the energy denominator terms of Eq. 2), with β values estimated in the range 1.2–1.6 Å^{-1} in the systems accessed by early experiments from the groups of Harriman [17] and Meade [18]. The early results of Barton [1–3] could not be understood in the framework of this single step-tunneling model [14–16]. The large β interactions in the computations arise because the neighboring π-orbital interactions occur "through space"—dominated by p-σ symmetry interactions [14–16]—and the donor–acceptor states are not in degeneracy with the bridge states (Fig. 2).

The distance and tunneling-energy dependence of H_{DA} and β suggested by Eqs. 1 and 2 is best probed in structures with well-defined donor–bridge–acceptor geometries. A second generation of DNA experiments was designed with more precisely positioned D and A groups. Many of these second-generation systems utilized DNA bases (guanines and their derivatives, for example) as redox partners. As the D and A are brought into near degeneracy with the bridging states, β is expected to decrease.

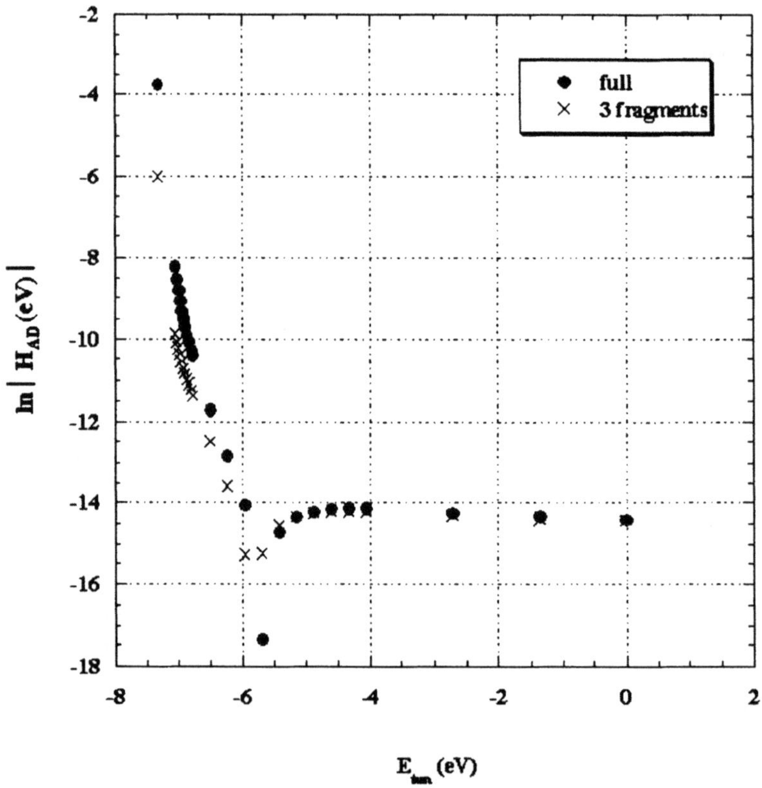

Fig. 2 Dependence of the electron donor–acceptor interaction H_{DA} in DNA on tunneling energy, E_{tun}. Near the filled or empty states of the bridge, the coupling is strong and changes rapidly as a function of E_{tun}. In the mid-gap region, the bridge-mediated coupling is nearly unchanged as tunneling energy varies. This simple picture is somewhat complicated by specific energies at which interferences cause sign change in the coupling. *Dots* correspond to the Hartree–Fock calculations for the entire system G A A G / C T T C while *crosses* correspond to divide-and-conquer calculations for the same system using two fragments, each involving three base pairs

A generic U-shape dependence of H_{DA} on tunneling energy is expected for tunneling through DNA. What more specific relationships link structure to couplings? Recent studies address this question by examining base–base interactions in a given strand and across strands. Much of the recent analysis is based on calculations of nearest-neighbor interactions and interpreting the coupling strengths in a McConnell-like Eq. 2 interpretation.

Bixon and Jortner, Rösch and Voityuk, Olofsson and Larsson, Berlin, Burin, Siebbeles, and Ratner, Orlandi and their coworkers [4, 8, 19–31] have explored the energetics and base–base interactions associated with hole and electron transfer in DNA base-pair stacks. They find nearest-neighbor coupling interactions of 0.03–0.14 eV on the same strand and 0.001–0.06 eV for cross-strand couplings (to partners not directly Watson–Crick base paired) [29]. Watson–Crick partners have couplings computed to be 0.03–0.05 eV. The virtual state energies of the bases in the calculations vary by 0.4–1.6 eV (depending on base and environment). Combining estimates of hopping interactions and base energetics allows an assessment of dominant coupling routes through the π-stack. These coupling matrix elements were usually derived from energy splittings through Koopmans' theorem. There is some ambiguity associated with the relative signs of matrix elements, so pathway interference effects may not be treated perfectly. Also absent from these calculations is the influence of the ribose phosphate backbone. Not withstanding these limitations, the energy parameters have allowed the construction of initial models for charge transport in DNA. Recent applications have been to kinetic schemes that span the mechanistic regimes from superexchange to hopping.

There are several concerns that are associated with McConnell-like analysis of tunneling that is based upon gas-phase base–base interaction calculations. (1) The dependence of the superexchange energy denominator on the reorganization energy λ is often neglected [13, 32, 33]. This effect is particularly important in the regime of rapid β variation with tunneling energy (i.e., $\beta<1$ Å^{-1}). (2) The influence of solvent on radical cation state energetics (especially the localizing effect of solvation) is usually neglected. Classical solvation and quantum delocalization energies are of comparable sizes in oxidized GC runs [34]. (3) The McConnell model is a "one orbital per site" model. For example, in the hole transfer regime, it neglects the contributions of the HOMO-1, HOMO-2, etc. virtual hole states of the bases to superexchange mediation. Since the density of π-electron states is high near the HOMO, this assumption could be particularly problematic.

3 Charge Transfer Between Native DNA Bases. Effects of Water Surroundings

Although theoretical suggestions of efficient charge transport through the DNA double strand were made 40 years ago [35], the mechanism of this process still remains the subject of active experimental and theoretical study. The actual dynamics of electrons and holes is much more complicated than the originally proposed band-like picture of charge motion along "π-pathways" provided by stacking interactions between heterocycles of base pairs [36]. In particular, theoretical analysis of clarifying steady-state measurements of the sequence dependence of the positive charge (hole) transfer efficiency [37–40] has led to the formulation of a phenomenological model [19, 40–42] that treats charge motion as a set of variable-range hops [41] between adjacent GC base pairs.

An obvious advantage of the model proposed is accurate predictions of both sequence and distance dependencies for the efficiency of hole transfer

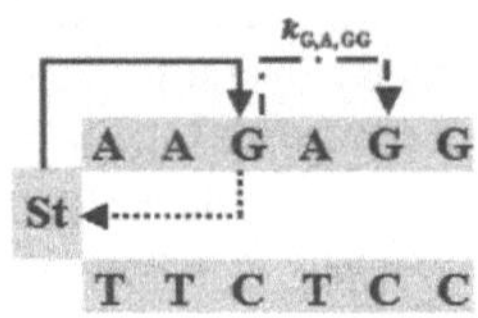

Fig. 3 Charge transfer in DNA hairpins after photoexcitation of stilbene linker (St) by a laser pulse [45]. A hole, first, undergoes a transition from photoexcited St to the adjacent GC pair as shown by the *solid arrow*. Then it can either hop to next GC pairs (*dot-dashed arrow*) or return to St with the subsequent electron–hole recombination (*dotted arrow*)

through stacks with various combinations of Watson–Crick base pairs [42]. This, in turn, provides reasonable evaluation of the distance scale for the propagation of a positive charge in DNA duplexes. Information needed for such predictions involves only the experimental data [43] on relative rates for hopping steps of different lengths. The significance of relative, but not absolute, rates for theoretical analysis of steady-state experiments is a direct consequence of two competitive decay channels existing for a positive charge at each step of the transport process, namely hole transfer between nearest-neighbor G sites through the AT bridge and irreversible side reaction of the cation G^+ with water.

Knowledge of relative hopping rates, however, is insufficient to decide how fast a hole generated in DNA can be transferred over a certain distance. To gain a deeper insight into this kinetic aspect of the problem, evaluations and direct measurements of the absolute value of the elementary hopping step rate are important. Early theoretical attempts to address the issue were based on the quantum mechanical nonadiabatic electron transfer theory [7]. According to the estimates made [44], the rate of hole hopping between G bases bridged by one AT pair is expected to exceed 10^9 s^{-1}. However, time-resolved experiments with DNA hairpins [45, 46] schematically depicted in Fig. 3 show that hole transfer for the process $G^+AGG \rightarrow GA(GG)^+$ is much slower than expected from nonadiabatic theory and proceeds with the rate $k_{G,A,GG} \cong 5 \cdot 10^7$ s^{-1}.

Recent theoretical investigations suggest that the discrepancy between theoretical and experimental values of the hopping rate is due to the underestimation of localization phenomena in the description of charge motion behavior. These phenomena can result from the coupling of electronic motion to vibrational dynamics of base pairs [47] and can also be caused by polarization of the molecule or solvent [34, 48]. In the latter case theoretically possible effects include:

(i) Self-trapping of a hole leading to the formation of a large polaron that spreads in length over 5–6 base pairs (i.e., about 17 Å) and moves together with the polarized surroundings along the stack of base pairs [48].

(ii) Competition between quantum delocalization of charges and their localization due to vibronic coupling or solvation forces, which causes the formation of rather compact hole states with the typical dimension less than 7 Å (the so-called small polaron) [34].

(iii) Response of solvent to charge motion favoring the adiabatic mechanism of hole transfer over the nonadiabatic mechanism.

The formation and transport properties of a large polaron in DNA are discussed in detail by Conwell in a separate chapter of this volume. Further information about the competition of quantum charge delocalization and their localization due to solvation forces can be found in Sect. 10.1. In Sect. 10.1 we also compare a theoretical description of localization/delocalization processes with an approach used to study large polaron formation. Here we focus on the theoretical framework appropriate for analysis of the influence of solvent polarization on charge transport. A convenient method to treat this effect is based on the combination of a tight-binding model for electronic motion and linear response theory for polarization of the water surroundings. To be more specific, let us consider a sequence

A ... A G A G A ... A G A A G
T ... T C T C T ... T C T T C

with N sites, in which hole transfer occurs between two GC pairs separated by one AT. Electronic motion in the sequence can be described in terms of the tight-binding Hamiltonian

$$H_e = -\Delta E \sum{}' n_i - b \sum_{i=1}^{N} \left(c_i^+ c_{i+1} + c_{i+1}^+ c_i \right), n_i = c_i^+ c_i. \qquad (4)$$

Here ΔE is the difference in ionization potentials of AT and GC base pairs, b is the transfer integral, c_i^+ and c_i are the creation and annihilation operators for a hole at the i-th site, respectively, index i labels DNA base pairs in the sequence, and the sum Σ' is taken over GC sites only.

To take into account the charge–solvent interaction, we introduce a multidimensional coordinate X_k that characterizes the polarization of water molecules around the k-th base pair. Then a standard linear response expansion enables one to express the interaction with the water environment as a function of X_k as

$$H_{\text{int}} = -\sum_i n_i X_i + \frac{1}{2} \sum_{ij} A_{ij} X_i X_j. \qquad (5)$$

According to Eq. 5, the interaction with water is determined by the site populations n_i, and by matrix **A** with elements A_{ij}, which can, in principle, be specified if solvation energies for oxidized states of G bases in DNA are known. So far calculations of solvation energies were performed assuming that polarization dynamics is slower than electronic motion [34]. Within this limit, migration of a hole initially localized on any G site can be viewed as the motion in the x direction proceeding along a certain one-dimensional

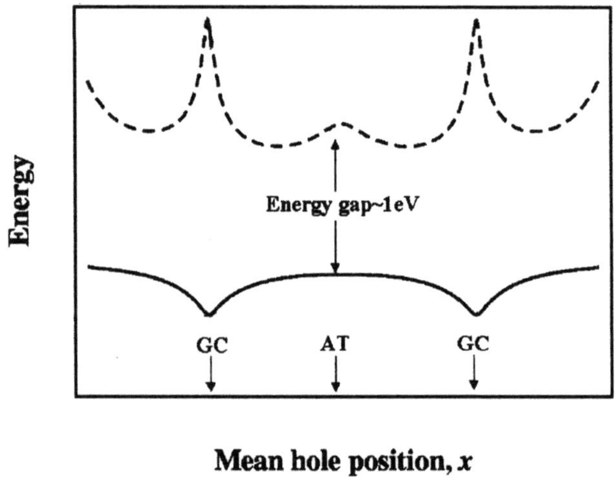

Fig. 4 Energy landscapes for ground- (*solid line*) and excited-state (*dashed line*) hole transfer in DNA. *Arrows* indicate mean positions of a hole matching the location of two GC sites bridged by a single AT pair. The width of the energy gap was estimated assuming that the contribution to polarization comes from nuclear motion

energy landscape. The profile of the landscape is determined by optimal values of the X coordinate, X_{opt}, at which the total energy of the system has minima. To construct this profile, the ground-state energy of the system should be minimized for each average position of the hole given by the coordinate x. This can be done using an iteration algorithm similar to the procedure proposed in [48] and a subsequent Monte Carlo search for X_{opt}.

The results obtained for different values of parameters ΔE and b for the particular sequence

A G A G A
T C T C T

show that the energy landscape for the ground-state charge transfer exhibits two symmetric global minima at x values corresponding to the presence of a hole on G bases, see Fig. 4. The barrier between two minima has the maximum height Δ_b at the point where x matches the location at the AT pair site flanked by GC neighbors. To transfer a hole from one G base to another, this barrier should be overcome due to fluctuations of molecular geometry in the local water environment. As a result, the rate of the elementary hopping step at temperature T can be approximated (in this adiabatic picture) by the Arrhenius-type expression

$$k \cong k_0 \exp\left(-\frac{\Delta_b}{k_B T}\right) \tag{6}$$

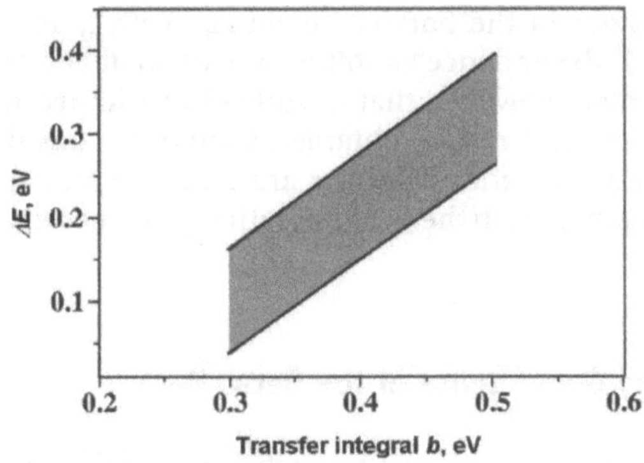

Fig. 5 Ranges of tight-binding parameters ΔE and b used in calculations of the hole transfer rate $k_{G,A,GG}$. The domain shown in *gray* include the values leading to $k_{G,A,GG}$ between 10^7 s^{-1} and 10^8 s^{-1}

with the pre-exponential factor k_0 being equal to the water orientational relaxation rate $1.2 \cdot 10^{11}$ s^{-1} [48] in accordance with general kinetic theory for over-barrier transitions [49].

The alternative mechanism of hole transfer might involve thermal excitation of the system. However, for temperatures below 300 K this seems to be unlikely, since numerical studies performed for typical values of parameters ΔE and b reveal the existence of the wide energy gap (~1 eV) between energy landscapes calculated for the ground and excited states (Fig. 4). Therefore "adiabatic" motion of a hole, which does not require excitation, remains the dominant channel for the transfer of a positive charge between the two G bases.

Evaluations of the barrier height Δ_b and estimations of the rate for hole transfer from Eq. 6 give a straightforward way to address the first problem. For the particular sequence shown in Fig. 3, this gives $k_{G,A,GG}$ varying from 10^7 s^{-1} to 10^8 s^{-1} depending on the values of parameters ΔE and b in the tight-binding Hamiltonian (Eq. 4), see Fig. 5. The obtained estimates were found to be consistent with the timescale of the process deduced from time-resolved experiments [45, 46]. For instance, if ΔE=0.4 eV and b=0.5 eV, the barrier height Δ_b is equal to 0.23 eV, and $k_{G,A,GG}=1.3 \cdot 10^7$ s^{-1}obtained from Eq. 6 turns out to be less than the experimental value by a factor of 4.

Similar calculations for bridges with different number, q, of AT base pairs between G sites allow conclusions about the behavior of the hole transfer rate as a function of the AT bridge length. The obtained dependence qualitatively is very similar to the length dependence predicted by nonadiabatic models [28] and observed in experiments [50]: the rate of hole transfer rapidly decreases as $q<3$ and remains almost constant for further elongation of the bridge. Explanations of this behavior offered by nonadiabatic and adiabatic theories are, however, different. For nonadiabatic transfer, the length dependence of the rate has been attributed to the transition from the tunneling mechanism of the elementary hopping step to thermal activation [28].

By contrast, changes in the barrier height Δ_b with q account for the same form of the length dependence as in the case of adiabatic transfer.

It should be noted, however, that in spite of qualitative agreement with experiment, the numerical results obtained should be considered with caution because the data on solvation energies are available only in the limit of slow polarization dynamics. Further investigations are necessary for going beyond this limit.

4 Tunneling Energy Dependence of the Decay Rate

The electronic coupling mediated by each bridge structure drops in proportion to the gap between the tunneling energy and the bridging state energies. This is a rapidly changing function when the two energies are nearly degenerate and a very slowly varying function far from resonance. As such, we anticipate a dramatic softening in the characteristic distance dependence of ET tunneling decay as the bridge states approach resonance with the donor. Indeed, this predicted tunneling energy dependence was seen in excited state ET in DNA [51]. The observed beta value in stilbene excited-state quenching by guanine oxidation yields the value of the falloff parameter β~0.7 Å^{-1}, while phenanthrene quenching by deazaguanine produces β~1.1 Å^{-1}. The tunneling energy in the latter case is shifted about 0.25 eV from the bridging states compared to the former (Fig. 6).

Theoretical analysis of electronic coupling in DNA [12] suggests that the limiting beta value—the values that would be accessed for donor and accep-

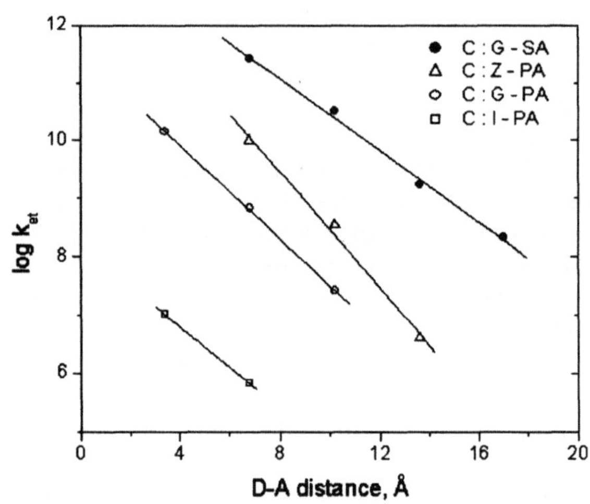

Fig. 6 The distance dependence of electron-transfer rates in DNA hairpins [51]. The acceptor is a photoexcited derivatized stilbene (SA) or phenanthrene (PA); the electron donor is guanine (G), deazaguanine (Z), or inosine (I). The decay is much more rapid in the Z–PA couple compared to the G–SA couple because the tunneling energy is further from the bridge states in the case of Z–PA

tor energies several electron volts removed from the bridging pi-electron states—is about 2 $Å^{-1}$. This large value contrasts strongly with the limiting values for alkanes, which are about half as large. Why are the two limiting values so different? In the far off-resonance regime (large energy denominators in Eq. 1), the main difference between the linear saturated bridge and the π-stack is the numerator of Eq. 2. In the case of the DNA π-stack, the V elements are roughly proportional to the overlap between the π orbitals that lie about 3.4 Å away from each other (these interaction strengths are of the order of tenths of eV), compared to the interaction between adjacent CC bonds (interactions of the order of eV). DNA ET systems are very special in this regard. Use of π-electron donor/acceptor species, combined with the π-electron character of the bridge, places these reactions in the near resonant regime, in which one has the opportunity to see tunneling energy effects. Note that this small tunneling energy gap regime is expected to be accessible for systems that are close to the tunneling/multistep hopping transition (*vide supra*).

5
DNA Conductivity and Structure

Neat DNA, because of its hybridization and complementarity behavior, is an obvious candidate as a synthon for nanostructures. Because naturally occurring DNA is a hetero-polymer, and because of all the structural disorder inherent in any DNA chain, such properties as coherence, transport, persistence length, and structure are expected to differ substantially among samples of DNA that have the same chemical composition. On the basis of this propensity for disorder, the statement that "this ignorance (of DNA transport) is justified by the complexity of the problem" is understandable [52].

Electrical conduction in DNA has become a highly contentious subfield of its own. The uncertainties here are even greater than those in the chemical measurements of charge transfer in DNA, partly because of the greater strand length generally involved, and partly because of the necessity for making contacts in measurement of DC conduction.

Electronic transport in DNA structures can be classified into three different subfields: neat DNA, doped DNA, and nanostructures using DNAs as templates. We will limit our remarks to the first two.

Doped DNA (by this we mean DNA strands whose electrical charge and electron count have been varied by transfer from electron donors or electron acceptors) can be thought of by analogy to doped electron crystals, but once again the narrow bandwidth and disorder characteristic of DNA make the problem more complex.

DNA as a templating material has become very important to nanostructures. The early reports involved DNA strands holding together arrays of coinage metal nanodots [53–56]. These hybrid materials are important in detection and sensing schemes; the readout of such sensing is generally through fluorescence or Raman spectra or surface plasmon measurements, although some electrical transduction schemes have also been utilized [57–59].

The measurement of charge transfer in DNA strands and charge transport in DNA conductance junctions both depend on the fundamental process of electron or hole motion from one end of the DNA strand to another. The measurements differ: in electron transfer, the rate constant k is measured, while in conductance junctions the conductance g is the observable. It is then reasonable to ask about the relationship of these observations. In the case of fully coherent single/step motion between the two ends, Nitzan [60] has developed some very nice rules of thumb relating k and g. This is based on the fact that both the Landauer formula for transport in coherent junctions and the Marcus–Hush–Jortner expressions for long-range electron transfer can be written as the product of electronic tunneling amplitude squared times an effective state density. Therefore, by solving for the amplitude squared, k and g can be related through their respective state densities. Nitzan has utilized some standard estimates for Franck–Condon factors and self-energy magnitudes to deduce the approximate results [60]

$$g\left(\Omega^{-1}\right) \sim 10^{-17} k\left(s^{-1}\right).$$

This means that for a rate constant of 10^{12} s^{-1}, one expects a conductance of the order 10^{-5} Ω^{-1}. Recently this concept has been extended to the case of incoherent transfer including the mechanism of variable-range hopping [61]. The relevance of this discussion for electron conductivity in DNA structures is that one would expect efficient long-range conduction only if long-range charge transfer is also efficient.

A second issue involves the transferability of parameters characterizing electron transfer in solution to conduction. The simplest conduction model, like the simplest electron-transfer model, is based on a tight-binding independent-electron picture (Hückel-like model) of electrons moving through a disordered set of levels corresponding to the HOMO or LUMO of the individual base pairs. Characteristic values for the energy differences among A, T, G, and C frontier orbitals and for the tunneling matrix elements, V, have been deduced experimentally and computationally, as discussed in Sect. 2. Once again, the relationship between conduction and rate constant can be based on the assumed transferability of these parameters from one measurement to another. Caution is needed, however, because DNA strands are easily distorted; in measurements made on, for example, a mica surface, the DNA strand is not necessarily in the same geometry as it would be in a DNA bundle, or in free solution. These distortion complexities will provide extra uncertainty in measuring and understanding conductance in DNA structures.

Because both charge transport and conduction are facilitated by electron motion between stacked base pairs, one expects that single-strand DNA should conduct far less well than double strand. Measurements both of rate constants and of conduction [62] indeed show that single-strand DNA is a far less capable transfer and transport agent for charge than is the duplex.

5.1 Neat DNA—Structure and Transport

Neat DNA is almost colorless. Calculations on single GC pairs give band gaps (associated with ionization potential/electron affinity differences) of the order of 10 eV. The unscreened matrix element, *V*, for perfect base pair alignment, is certainly no larger than 0.6 eV [63]. Taken together, these data suggest that neat DNA, even in the absence of distortion, should be a large-gap, narrow-band semiconductor.

One striking result [64] involves making a small change in regularity: when one GC pair in a strand is simply inverted to CG, the local defect CG pair drops 0.6 eV below its previous position—that drop is 15 times the calculated bandwidth. This electrostatic stabilization means that the defect level is far from the conducting delocalized states, and corresponds to Anderson-type localization. With Anderson localization of this depth, the conductance is expected to decay exponentially. Indeed, exponential decay of conductance has been discussed in a number of measurements both on λ-DNA and on poly(GC) sequences [65–67].

In λ-DNA, with its naturally occurring aperiodic structure, one deals with a strongly disordered tight-binding band. In such bands charge carriers are expected to be fully localized, and coherent electronic motion will decay exponentially with chain length. Activated motion in the form of polaronic hopping has been discussed extensively (see, e.g., [66, 67]) and the disordered tight-binding picture suggests strongly that polaron-type motion should become dominant except at very low temperatures. Because such polaronic motion is essentially hopping type, exponential decay of conductance with length is no longer expected in the polaron mechanism.

Although these arguments concerning disorder and localization seem strong, they are not universally accepted. For example, some measurements [68] using rhenium–carbon electrodes suggest that λ-DNA exhibits quantized-conductance, proximity-effect superconductivity below 1 K, and the power-law temperature dependence of a Luttinger liquid. This work discusses transport in terms of low-temperature phase coherence over a few hundred nanometers, and says that DNA is metallic throughout a broad temperature range up to and including room temperature. These observations conflict directly with most other measurements, although a very important early measurement of λ-DNA subjected to intense radiation by 20–300 eV electrons does show that ropes of λ-DNA under these conditions exhibit ohmic, metallic conduction [69]. These authors suggest that the details of the experiments, including contact resistances, could well account for quantitative disagreements in resistivity values. They also suggest that DNA might be an insulator, a semiconductor or a metal depending on "the way one arranges the molecules on the structure in which one wants them to function in a certain way."

The well-known helicoidal distortions [70] that even two adjacent base pairs can suffer (role, twist, slide, etc.) suggest that, indeed, the electron/vibrational coupling should be large in DNA, and therefore that polaron formation should be facile. Moreover, the fact that DNA is a polyelectrolyte suggests that the ionic motion of the counterions should provide an extra term in the reorganization energy that will also affect charge motion. A recent extensive MD simulation has been used as the basis for arguing that the ions will very strongly control transfer along DNA strands [71].

Polaronic motions would give interesting temperature dependence of DNA transport. The important early measurements in Delft on transport in bundles of GC strands 30-base-pairs long suggested that the band gap actually increases with increase in temperature [72]. This can be explained on the basis of the temperature dependence of the tunneling matrix elements, if the effective tunneling matrix elements break the Condon approximation and are dependent on geometry [23]. Both electronic structure calculations [22, 24] and simple reasoning suggest that the 36 degree pitch between base pairs in the standard DNA double helix will not optimize the overlap—if DNAs are allowed to twist, both the overlap and the local matrix element should increase as the base pairs come closer to superimposition.

5.2 Electrical Transport. Measurements and Interpretation

In any measurement of dc electrical transport, it is necessary to have electrodes. In nanostructures, the contacts between the conducting structure and the electrodes can (and generally do) completely dominate the transport response. Therefore, it can be helpful to measure the conduction in an electrodeless geometry. Several investigations of the ac response of the transport in DNA stacks have been reported. Earlier work by Warman and collaborators [73] demonstrated that B-form DNA transports charge substantially better than A-form. These experiments also suggested that DNA does not have high charge mobility and that transport is strongly dependent upon the amount of water. Indeed, the interpretation of the experiments left some doubt as to whether ionic or electronic transport was measured.

More recent measurements by Gruner and collaborators [74] report conductivities in DNA in the frequency range 12–100 GHz. They observe that dry DNA has a conductivity of 0.1 S/cm, increasing in the buffer solution to 2 S/cm. They deduce a resistance for a 17-μm double-strand structure as 10^{10} Ω, and for a 600-nm double strand as roughly $5 \cdot 10^8$ Ω. The observed temperature dependence is reminiscent of both small polaron transport and the transition from quantum tunneling at low temperatures to activation at high temperatures—the activation energies in the high-temperature regime (above roughly 250 K) approximate to 0.33 eV, and for temperatures below roughly 200 K, the conduction is weakly temperature dependent. They also observe only weak frequency dependence. The conductivities are comparable to those of the one-dimensional ionic conductor potassium hollandite. If the conduction is fit to an activated form, the prefactor (σ_0) is roughly 200 S/cm, comparable to several organic semiconductors. They also suggest possible ionic conductivity.

Other electrodeless measurements can be made using electrostatic force microscopy, from which it has been deduced [75, 76] that λ-DNA exhibits very poor conductance—a conductivity of less than 10^{-16} S/cm was suggested. Indeed, no evidence for significant conduction was seen either in λ-DNA or in poly(GC).

More traditional transport measurements involving either two electrodes or an added gate electrode have been reported. These various publications report DNA acting as an insulator, a semiconductor, a metal, or a proximity superconductor.

This very broad range of conduction reports (more than 12 orders of magnitude) is dismaying. As indicated previously, some of it can be understood on the basis of the different structures that are in fact being interrogated: in addition to the differences between dry and wet DNA, λ-DNA and poly(GC), A form and B form, solvated and dried, free standing and supported, there are substantial differences in the nature of the electrode/DNA contacts.

A rapid survey of some of the more recent publications seems to point to DNA indeed being a very broad-gap semiconductor or insulator for lengths exceeding a few hundred base pairs. This conclusion was reached in references [74] and [75] on the basis of electrostatic imaging, in reference [77] using conducting atomic force microscopy on λ-DNA, in [78] using 300-nm DNA strands both of poly(GC) and of λ-DNA, and in [52] for λ-DNA supported on mica.

Dekker and collaborators have published two numbers for DNA conduction: in Ref. [65] they conclude that on the 100-nm-length scale DNA is an insulator, while on the 10-nm-length scale (see [72]) poly(GC) is suggested to be semiconducting.

Other reports of semiconductivity also exist in the literature. These include some important studies of doped stacks. Reference [78] uses oxygen hole doping in poly(GC) to control the conductance. The authors cite the increase of conductivity by several orders of magnitude upon oxygen hole doping to suggest that poly(GC) is a p-type semiconductor, and that this control of conductance argues strongly against any ionic contribution to the conductivity. They also suggest that poly(AT) is an n-type semiconductor. In [79], a three-electrode configuration is again used, although the DNA is clumped rather than linear. The assignment of poly(AT) as n-type and poly(GC) to p-type is again found, and polaron hopping is invoked as the primary mechanism of charge transport. In [77], a three-probe nanotube-based AFM measurement is reported—the authors observe regular jumps in the source/drain current with a spacing of 80 MeV, and a voltage gap decreasing with increased gate voltage. According to [64], poly(GC) shows higher conductivity than poly(AT), and conducting AFM measurements show that the resistance increases exponentially with length.

In the light of these reports, the suggestions that DNA behaves either as a metal [68, 81] or as a proximity-effect superconductor [68] seem striking. Part of the issue certainly has to do with the electrodes—in [68], rhenium/carbon electrodes were used. The observations reported in [68] (quantized conductance, metallic conductivity, power-law temperature dependence) are those expected of coherent molecular wires in the ohmic regime, where no electrode effects occur. It is possible that the rhenium/carbon electrodes used in this study indeed made ohmic contact; if true, this would be a striking advance. The pioneering measurements of [69] were done with λ-DNA on a supporting grid—the DNA was treated with a high flux of electrons in the 20–300 eV range. It has been suggested that, since in this range electron-induced damage is widely observed in biological samples [82], it is possible that the actual substance whose conductance is being measured is no longer the native DNA from which the sample was originally prepared.

Although measurements continue to be made, it seems quite clear that native DNA is certainly not the band-type conductor first suggested by Eley and Spivey [35]. Disorder effects, polaron coupling, and the narrow band nature of the transport all suggest that λ-DNA, or any native DNA, should be an insulator at long length scales. Quoted measurements indeed support this: conductivity of less than 10^{-15} S/cm [76], or $1/g > 10^{13}\ \Omega$ for lengths ~40 nm [65], $\sim 4 \cdot 10^{10}\ \Omega$ for poly(GC) [77], or $>10^{12}\ \Omega$ for 1.8-μm DNA [52].

It is perhaps worth remembering that proper "band alignment" can indeed cause perfect conduction even in molecular wires where the delocalization is relatively weak. This can be seen formally starting from the Landauer formula [83], and is also fairly clear conceptually: if injection occurs precisely on resonance, and if there are no dissipative mechanisms (as in Landauer conductance), then the scattering probability through the molecular junction will be unity, and quantized conductance will be observed. This has been very recently seen in gated tunneling junctions with small molecules [84, 85]. DNA measurements using a gate electrode are still quite rare [75, 76], but these measurements, in which the number of charge carriers can at least to some extent be controlled, should be important in understanding the mechanistic behavior of DNA. More such measurements will certainly be forthcoming.

Altogether, the preponderance of direct conduction measurements seems to indicate that DNA is indeed a wide-gap semiconductor or insulator, that the charges are localized over a few base pairs or less at ambient temperatures, that polaron-type effects should be important in long-range charge transport, that over certain ranges the conductance decreases exponentially with length, that ions make no substantive contribution to the transport, and that expanding the disorder by adsorption onto surfaces or supercoiling reduces the transport yet further.

While many of the issues remain controversial, it is clear that electrode contact will dominate transport under certain conditions[1], and that the mechanistic issues involved once the electrode problems are understood relate closely to the same mechanistic issues involved in electron transfer experiments. While band structure calculations are very helpful for under-

[1] For truly one-dimensional situations, screening at the electrodes will be very much reduced, so that, after extensive charging, higher conductivity might be expected (J. Xu, personal communications).

standing the magnitude of the bare (no vibronic coupling) electronic structure, the strong propensity for coupling to vibrations, as well as the ionic environment of the polyelectrolyte and the relative floppiness of DNA structures, suggest that vibronic transport should indeed be dominant over lengths corresponding to more than ten or so base pairs. Some of the most striking effects seen in long-range electron transfer measurements have not yet been observed in transport, basically because the measurements are still not terribly reliable or reproducible. Such effects include thermally induced hopping and the coherent/incoherent transition.

In sum, then, although fascinating reports of conductivity in DNA structures have been published, and some general structure/function motifs have become clear, difficulties with reproducibility of experimental data and with appropriate interfaces between nanoscale DNA structures and macroscopic electrodes have limited the accuracy with which DNA charge transport can be measured, and the depth in which it can be understood. This remains an active area, and (especially given DNA's very powerful presence as the synthetic component of nanostructures) it is one that will almost certainly be more clearly elucidated in the near future.

6 Vibronic Coupling, Reorganization Energies, and Ionic Gating

In Sect. 4, we have already indicated that the electronic energy of a nonequilibrium "transition state" species defines the energy mismatch between donor/acceptor and bridging species. This energy mismatch impacts the coupling matrix element. The charge distribution change between reactants and products, as well as the polarization characteristics of the medium, determine the activation free energy for the process.

6.1 Reorganization Energy and DNA Electron Transfer

The classical Marcus reorganization energy λ is the inertial component of the system's electrostatic energy. As such, it is computed—within a dielectric continuum approximation—in the following manner. First, the electrostatic energy of the product associated with the electron shift is computed assuming a uniform high-frequency dielectric constant (ε=2). Second, the electrostatic energy is computed using a low-frequency dielectric inside the DNA (ε=2 or 4) and a larger low-frequency dielectric for the solvent (ε=80). The difference between the two electrostatic energies gives λ [33]. Assuming spherical donor and acceptor groups, the classical Marcus expression for the reorganization energy is obtained. The corresponding classical activation free energy is

$$E_{\mathrm{act}} = \left(\Delta G^0 + \lambda\right)^2 / 4\lambda, \tag{7}$$

where ΔG^0 is the driving force.

Analysis of reorganization energies with spherical (or more involved) models predicts a rapid increase in λ with distance at short distances, and a weak distance dependence at longer distances. Since the rate depends exponentially on the activation free energy, changes in λ with distance can have an extremely large influence on rates. Indeed, Tavernier and Fayer [32] were the first to point out that, if one assumes single-step donor-to-acceptor tunneling in DNA electron transfer, the observed distance dependence can be explained by the increase in λ, without introducing any distance dependence in the coupling matrix element! Dividing space into spherical regions—each with an assigned dielectric constant—they predicted a doubling of λ as the number of intervening bases changes from zero to four.

Tong, Kurnikov, and Beratan [12] used a finite-difference method to compute reorganization energies. This method accommodates arbitrarily shaped molecules. The authors investigated changes in λ for guanine-to-guanine hole transfer and found similarly large changes in λ with distance. In this study, the authors also examined the distance dependence of the coupling matrix element for single-step hole transfer from GC to GC through ATs. The increase in reorganization energy (slowing the rate) and decrease of H_{DA} (also slowing the rate), in combination, would predict a more rapid decrease in overall rate as a function of distance than is seen in any experiments.

6.2 Ion-Coupled Electron Transfer

The above discussion assumes that transfer rate is controlled by one or more vibronically coupled ET events. That is, the reaction coordinate for the process brings donor and acceptor states into quasi-degeneracy, at which point the ET event occurs. The rate depends on both the likelihood of reaching the activated complex and the tunneling probability in that (those) complex(es). This standard formulation assumes that there are no slow "gating" processes that are rate limiting. Gated events are well known in electron transfer. Some small-molecule intermolecular ET rates are limited by counterion motion [6, 33], and macromolecule ET rates can be gated by conformational changes. Recent simulations and experiments of Landman, Schuster, and coworkers [71] point to the possibility that cation motion could gate ET in DNA. Their simulations suggest that counterion positions around the double helix define hole transfer "effective" and "disabled" configurations. Experimentally, the neutralized DNA with methyl phosphonates is a less efficient hole transporter than λ-DNA. Whether this decreased yield arises from a static change in radical-cation energetics or a true dynamical gating associated with counterion motion remains an open question.

6.3 Backbone vs Base Pair Tunneling Mediation

The early studies of Risser, Beratan, and Meade [13] suggested that, at short distances (especially for backbone-tethered donor–acceptor pairs), backbone mediation could dominate the coupling interactions. Later studies by Priyadarshy, Tong, Kurnikov, Risser, and Beratan showed that—for π-stacked donor–acceptor pairs—base interactions clearly dominate the bridge-mediated coupling interactions [12, 16].

6.4 The Condon Approximation in DNA Electron Transfer

Since the ET coupling depends upon an energy denominator, the coupling element will be hypersensitive to energy changes when this denominator is small and nearly insensitive to the denominator when the energy is large. That is, when the tunneling energy E_{tun} in Eq. 1 approaches the energies of the bridging states E_{Bi}, small shifts in the E_{tun} values can have a large effect on the mediated coupling. One formulation of the Condon approximation fixes the tunneling energy at a value determined by the reaction free energy and reorganization energy. Comparison of the rates obtained within the Condon approximation and those calculated with inclusion of a tunneling-energy-dependent coupling indicate that the Condon approximation introduces modest (~20%) errors to the rate [12] (for more details, see sect. 9).

7 Timescales and Traps

Because electron transfer is a rate phenomenon, considerations of timescale are unavoidable in the modeling and understanding of ET reactions. Indeed, even in simple polaron theory (vibronic theory for electron transfer between two sites), there are several timescales including

$$\frac{\hbar}{k_B T}, \frac{\hbar}{V_{rp}}, \frac{\hbar}{\lambda}, \frac{\hbar}{\sqrt{k_B T \lambda}}, \frac{1}{\omega_v}, \frac{\hbar\lambda}{(\Delta G_0 + \lambda)^2} \tag{8}$$

Generally, reactions are considered to be nonadiabatic when the effective splitting, $2V_{rp}$, is smaller than the thermal energy. This is, however, an inexact prediction—polaron theory provides a more complete set of demarcations between the two limits [86].

An important concept is that of the contact time. This is originally defined by Landauer and Buttiker [83, 87] as the so-called tunneling time, and is best envisioned as the time during which the tunneling charge actually contacts the vibrational bath. In the Landauer–Buttiker formulation, this time is given as

$$\tau_{\mathrm{LB}} = \left(\frac{m}{2(U-E)}\right)^{1/2} D. \tag{9}$$

Here the variables m, U, E, and D are the mass of the tunneling particle, the height of the rectangular barrier through which it is tunneling, the tunneling energy, and the width of the barrier, respectively. Notice that τ_{LB} becomes longer as the particle becomes heavier or the thickness of the barrier increases—that is perfectly reasonable. Notice also, though, that as the barrier height increases, the time of contact actually decreases. This shows that the tunneling time has nothing whatever to do with the rate time, since that is clearly slower through a high barrier than through a low one. The tunneling time is simply the time during which the electron is in contact with the bridge, and it indeed decreases as the barrier height grows.

The Landauer–Buttiker time was derived for tunneling through a rectangular barrier. In the case of a series of electronic levels, such as occur for DNA in the models of Fig. 1, one can generalize the Landauer–Buttiker argument to a molecular orbital representation. In the limiting case that is appropriate to most DNA-type problems, the time in Eq. 10 holds [88]

$$\tau_{\mathrm{LB-M}} = \frac{\hbar N}{\Delta E_b}. \tag{10}$$

Here, N is the number of sites through which the electron or hole tunnels, and ΔE_b is the energy barrier between the injection energy and the barrier height. In accord with physical intuition, this suggests that longer barriers will take longer times for tunneling. As in the Landauer–Buttiker result, higher barriers give shorter tunneling times. For characteristic tunneling barriers of the order of 0.5 eV, and for a five-site bridge, this gives a tunneling time of the order of 5 fs. This is shorter than characteristic vibration frequencies, and suggests that strong electron or hole trapping by the vibrations to form a polaron is improbable. As the bridge length gets longer or the gap gets smaller, the tunneling time will approach the frequencies of characteristic vibrations. Under these conditions, one expects strong inelastic effects, possible breakdown of the Condon approximation (see Sect. 9), and substantial corrections to the simple superexchange, effective two-site model.

The tunneling time estimate based on Eq. 8 also explains why thermally induced hopping may be significant for long strands of ATs between GC sites: the tunneling time can then become long, so that the vibronic interaction on the AT bridge is strong and the particle can indeed undergo localization.

These evaluations can be helpful in understanding when superexchange pictures are inadequate, and when it is necessary to consider vibronic coupling on the bridge, adiabatic transfer, and the dynamics (as opposed to single geometry behavior) of the mixing elements that modulate tunneling along DNA strands.

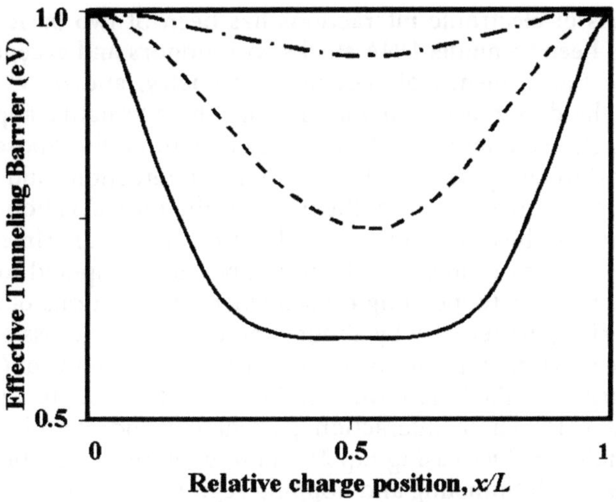

Fig. 7 Reorganization of tunneling barrier for the transfer of electrons, interacting with vibrations in different regimes of slow, intermediate, and fast tunneling [90]. The *dot-dash, dash,* and *solid lines* refer (respectively) to $\omega_v\tau_t$=0.1, 1.0, and 10.0, see text in Sect. 7

One interesting issue has to do with the actual time of contact between the tunneling electron (or tunneling hole) and the vibrational structures along the bridge with which it is in contact. In the simple Landauer theory for coherent tunneling, such interactions are unimportant. Extensive calculations have shown that elastic energy exchange from polarization of the environment can lower the effective energy barrier and therefore increase the rate [89, 90]. Figure 7, taken from a semiclassical calculation [90], shows the effective barrier for the electron as a function of tunneling charge position between donor and acceptor in a model of the vibronic coupling reaction. Note that the barrier becomes smaller as the dimensionless product of the tunneling time τ_t and the vibrational frequency ω_v increases from 0.1 to 10. This is due to the successful reorganization of the vibrational mode of the molecule that adapts itself to the charged state.

8 Particular Site Combinations and Potential Well Depths

Specific theoretical studies of DNA structure, transport kinetics, and reaction dynamics have been diverse. Analysis has included: (1) master equation schemes to analyze competing mechanisms (e.g., hopping vs tunneling), (2) models for interactions with the bath (e.g., Redfield theory) to explore dephasing mechanisms, (3) electrostatic analysis of DNA solvation energetics, molecular dynamics studies of ion gated ET, and (4) approximate solutions of the Schrödinger equation to probe electronic interactions among bases [19–31].

The analysis of electronic interactions has been of two basic kinds. One approach has been to model DNA stacks with donors and acceptors bound, and to probe the influence of sequence, geometry, and σ- vs π-coupling pathways on the donor–acceptor interaction. The alternative approach is to explore interactions in model π-stacks (usually without the ribose phosphate backbone) and to compute base–base coupling interactions and relative energies (usually within Koopmans' theorem) of the radical cation base states. Comprehensive studies have been carried out for base-pair trimers. Nearest-neighbor interactions in idealized B-form DNA are computed to vary by as much as a factor of 5 (depending on sequence), interactions of a base with its Watson–Crick partner vary by about a factor of 2, and cross-strand interactions between bases in positions i and $i\pm1$ vary by a factor of 60 [29]. Virtual state energies of the base radical cations vary from 0.4 to 1.6 eV, as discussed above [21]. These interaction parameters and energies have been used to estimate β values (using Eq. 2) and to explore ET kinetics in models that can access both tunneling and hopping regimes [27, 28].

While base-pair trimer and extended-chain calculations provide important insights, it is clear that the geometric and structural variations of "real" DNA make quantitative estimates difficult. The electronic structure calculations are most relevant to "virtual" states of the DNA, that is states that gain or lose an electron without relaxation of the nuclei. In the case of bridge oxidation, this assumption will be far from perfect. Even in the case of superexchange, electronic polarizability of the surrounding DNA backbone and solvent is absent in current considerations and can shift the virtual state energetics. Overall, solvation (including counterion motion), intramolecular relaxation of the oxidized bridge states, fluctuations of geometries and interaction energies, and mechanistic transitions between tunneling and hopping make the DNA ET problem very challenging to model quantitatively. Recent studies of Bixon and Jortner [27] aimed at linking the thermally induced hopping and superexchange models using the electronic structure calculations of the modest base-pair stacks discussed above. Their results were satisfactory from a qualitative point of view (predicting the distance at which the mechanism switches from single-step tunneling to multistep hopping), but disappointing in terms of agreement with product yield data and the elementary rate inferred from the ET kinetics.

In the superexchange regime, the tunneling may be mediated by the DNA bases, by the backbone, or by the solvent. Theoretical (INDO) studies by Cave and coworkers predicted a β of about 2.0 Å^{-1} for water, and they estimated (using a limited set of water configurations) that the ab initio value would fall in the range of 1.5–1.8 Å^{-1} [91]. The experiments of Gray and coworkers have demonstrated [92] that β for ET from ruthenium complex excited states to ferric ions in acidic glassy water at 77 K is 1.6–1.7 Å^{-1}. At a tunneling distance of two base pairs (6.8 Å), the coupling decays by a value not much smaller than that observed for π-stack mediation. Moreover, certain donor–acceptor geometries favor tunneling mediation by the ribose phosphate backbone. Extended-Hückel calculations [13] have indicated that, for systems with backbone-attached donors and acceptors, the apparent sign of β can change as a function of distance (that is, rates accelerate rather than slow as distance grows for cross-strand attachment geometries).

9
Breakdown of the Condon Approximation

The familiar equilibrium form of the Marcus–Hush–Jortner equation is simply

$$k_{\mathrm{ET}} = \frac{2\pi}{\hbar} < V^2 > (\mathrm{DWFC}). \tag{11}$$

Here the two factors on the right are, respectively, the average squared mixing (electronic structure tunneling interaction) between donor and acceptor and the density of states weighted Franck–Condon factor (DWFC). To derive this form, it is necessary to assume that the mixing matrix element is independent of the vibrational modes that contribute to (DWFC). This assumption is a generalization of the Condon approximation used for optical spectroscopy, and in the electron transfer literature is generally referred to as the Condon approximation. The physical meaning of this approximation is simple: the molecules of interest in fact do undergo vibrational excursions, but the matrix element that describes the mixing between donor and acceptor should be nearly independent of this motion in order for the rate equation, in the form of Eq. 11, to hold.

DNA is a floppy molecule. Although its persistence length can approach 100 base pairs, natural DNA exhibits supercoiling behavior, and its wrapping around nucleosomes in the chromatin structure is crucial for its biological function. Analysis of DNA bending and bendability, both experimental and theoretical (often using the helicoidal parameters that describe motions of the bases as rigid structures), has been widespread. Molecular dynamics simulations by a number of workers have demonstrated clearly that the DNA strand is quite floppy in aqueous solution at room temperature [93, 94].

The floppiness of the DNA strand might suggest that the Condon approximation fails, and that the mixing between local sites should be dependent upon the geometry of the strand. While this case is nowhere near so clear as it is in (say) biphenyl, nevertheless the floppiness might be expected to change the overlaps and therefore the matrix elements between local base pairs. In a structure like biphenyl [95], the twisting motion around the bridging single bond changes the character from quinoid to localized, and therefore completely changes the nature as well as the strength of the mixing between the two rings. Similar twisting motions provide important resistivity paths in molecular metals [96, 97]. One can legitimately ask if such effects are important in DNA.

Formally, if the mixing is dependent upon the geometry, the form of Eq. 11 will no longer hold. One can show [98] that, if most of the modes of vibration do not affect the matrix element substantially, one can derive a form that is reminiscent of Eq. 11. This can be done by separating the vibrations into a set denoted by v that does modulate the matrix elements, and another set, labeled w, that does not. Under these conditions, the rate constant becomes

$$k=\frac{1}{\hbar^2}\sum_{\mathrm{v}} P_{\mathrm{v}}\int dE\int dt < V(t)V(0) > \rho_{\mathrm{FCv}}(E)\exp\left(\frac{it(E_{\mathrm{v}}-E)}{\hbar}\right) \tag{12}$$

Here the average describes the correlation of the mixing terms, P_{v} is the probability of the system to be found in the vibrational state v, and the generalized Franck–Condon factor is defined by

$$\rho_{\mathrm{FCv}}(E)=\sum_{\mathrm{w}} S_{\mathrm{vw}}\delta(E-E_{\mathrm{w}}) \tag{13}$$

with S_{vw} being the Franck–Condon overlap factor. When the matrix element becomes time independent, Eq. 12 collapses to Eq. 11. More generally, however, the modulating motions of the mixing will affect this integral, calling for generalized treatment. One possibility is to expand in moments; another is simply to do a numerical investigation.

In the case of DNA, Troisi and Orlandi [26] have completed an integrated molecular dynamics/electronic structure study that examines precisely the issue of the dependence of the matrix element on geometry. These investigators studied a ten-base-pair double helical strand, whose interior contained two GC base pairs separated by four AT base pairs. They placed $\begin{matrix} \mathrm{G} & \mathrm{A} & \mathrm{A} & \mathrm{A} & \mathrm{A} & \mathrm{G} \\ \mathrm{C} & \mathrm{T} & \mathrm{T} & \mathrm{T} & \mathrm{T} & \mathrm{C} \end{matrix}$ sequence of these oligomeric DNAs in aqueous solution, completed molecular dynamics simulations, and then calculated the matrix elements as a function of time.

Some of their results are shown in Table 1. These results allow several conclusions. First, the matrix elements for two GC pairs separated by four AT pairs are all quite small, small enough to justify easily the use of nonadiabatic models to describe the electronic motion. Secondly, we notice that the average coupling is smaller, by somewhere between a factor of 4 and a factor of 16, than the mean square coupling. The actual dynamical simulation shows that the effective mixing changes its sign as well as its magnitude, and that it oscillates at an amplitude of roughly one wave number around zero. The final rate constant, from Table 1, will be enhanced by two orders of magnitude by considering the RMS mixing rather than the average mixing.

Table 1 Time-averaged base-pair couplings (in cm^{-1}) and their variance σ

Couple	V_0	σ
G3-A4	−1842	1350
A4-A5	−625	928
A5-A6	−150	715
A6-A7	−1442	663
A7-G8	148	620

Generalization of this concept, on a formal basis, has recently been presented [98].

These results suggest that the Franck-Condon approximation actually deals badly with DNA strands, especially for weak, long-range superexchange-type hopping. Looking at the animations of the molecular dynamics simulations makes it clear why this is true: it is not that any given frequency is important, but that the stochastic average of all the frequencies results in near cancellation (the average term is small), while large fluctuations still remain. This is manifest in the Fourier transform of the correlation function, which maximizes at a frequency of zero corresponding to the pure stochastic modulation.

The breakdown of the Condon approximation can actually lead to gating phenomena if particular angles are very much favored. It can also lead to situations in which the Marcus–Hush–Jortner formula must be generalized to deal with the Condon breakdown, and the results can be quantitatively important. DNA seems to be a situation in which Condon breakdown is striking, and this should be kept in mind when comparing adiabatic and nonadiabatic transport mechanisms.

10 Fluctuations and Injection

10.1 Radical Cation Delocalization and Energetics

We turn now to the nature of the oxidized states of guanine and multiguanine-containing sequences. Hartree–Fock/Koopmans' theorem-based quantum calculations of GC stacks indicate a change in the stability of the hole by ~0.7 eV as the chain grows from one to four base pairs [99]. Experiments, however, indicate that these hole states differ by no more than ~0.1 eV. Kurnikov, Tong, Madrid, and Beratan [34] explained this fact by a competition between the (localizing) solvation forces and the (delocalizing) electronic interactions among bases. As has been mentioned in Sect. 3, their estimations suggest that holes are delocalized over no more than three base pairs (the compact hole state with the typical dimension less than 7 Å).

An alternative phenomenological approach taken by Conwell, Basko, and Rakhmanova [48, 67] is to model the DNA as a one-dimensional chain with nearest-neighbor interactions among units of the chain. There is one effective orbital per unit. They assume that the sites are linked by harmonic springs. The nearest-neighbor interactions are taken to vary linearly with distance between the neighboring units (this model is motivated by the Su–Schrieffer–Heeger Hamiltonian [100] developed to describe the electronic structure of polyacetylene). Here, localizing forces are provided by chain relaxation around the hole. While highly parameterized by the effective hopping interactions, spring constants, and vibronic interactions, the model predicts cations/anions extended over five to seven base pairs with decreased hole binding energies. Both models with vibronic interaction limited to intramolecular DNA interactions and models that take the vibronic coupling as arising from solvation forces predict similar hole delocalization lengths.

10.2
Composite Hopping-Injection-Tunneling Models

The predicted strong increase of reorganization energy and rapid decrease of superexchange couplings (H_{DA}) with distance in DNA seem to eliminate the possibility of single-step long-distance ET in DNA. For intermediate distances (more than about three base pairs) motion likely has a significant hopping component. Bixon and Jortner [4, 27] and independently Berlin, Burin, and Ratner [28, 40, 42] have constructed quantitative models to describe these "hybrid" kinetic processes (see Fig. 8). The models are reminiscent of carrier transport models in organic semiconductors [101], and key parameters involve injection free energies, as well as short-range superex-

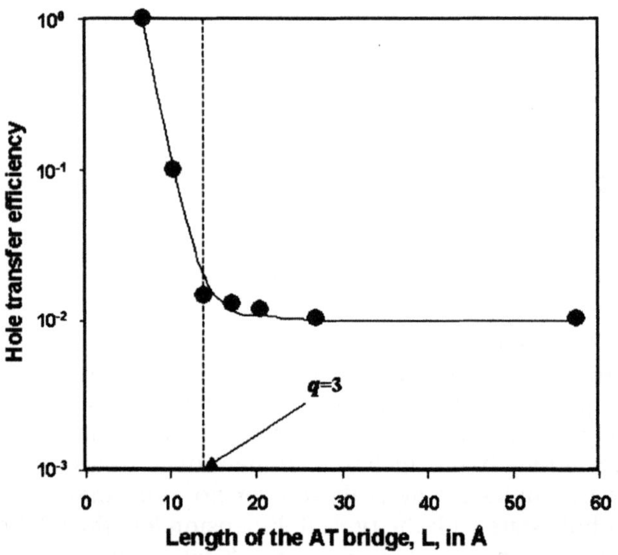

Fig. 8 Efficiency of hole transfer from G^+ to GGG across AT bridges of various lengths, L, [28]. Points correspond to the experimental data of Giese et al. [50] on the damage ratio for sequences $\begin{matrix}G\\C\end{matrix}\begin{pmatrix}T\\A\end{pmatrix}_q\begin{matrix}G\\C\end{matrix}\ \begin{matrix}GG\\CC\end{matrix}$ as normalized to the value of this ratio for the bridge with one AT pair (q=1). For each sequence, the damage ratio P_G/P_{GGG} is defined in terms of the measured time-independent yields P_G and P_{GGG} for the products formed in the reactions of water with G_j^+ and $(GGG)^+$, respectively. The length dependence of the hole transfer efficiency [28] is shown by the *solid line*. The intersection of the *dotted line* with the horizontal axis shown by the *arrow* indicates the bridge length and the number of AT pairs, at which the rates of tunneling and thermally activated transitions become equal

change interactions. While these models are satisfactory in their qualitative description of transport [28], entirely satisfactory quantitative descriptions await researchers in the future.

11 Concluding Remarks

In the absence of dynamic and static disorder, all partially filled band systems would exhibit coherent transport over long distances. With static and dynamic disorder, the modulation of the simple molecular orbital or band structure by nuclear effects entirely dominates transport. This is clear both in the Kubo linear response formulation of conductivity and in the Marcus–Hush–Jortner formulation of ET rates. The DNA systems are remarkable for the different kinds of disorder they exhibit: in addition to the ordinary static and dynamic disorder expected in any soft material, DNA has the covalent disorder arising from the choice of A, T, G, or C at each substitution base site along the backbone. Additionally, DNA has the characteristic orientational and metric (helicoidal) disorder parameters arising from the fundamental motif of electron motion along the π-stack.

The extensive disorder in any DNA structure leads to a myriad of mechanisms for electron transfer in rate constant measurements, and for electron or hole transport in conduction measurements. The major difficulties of understanding electron motion mechanisms in DNA are then of two sorts. First, the different kinds of disorder make reproducible measurements difficult to obtain. Second, the different sorts of interactions (Coulombic, vibronic, polarization, dynamical relaxation) make well-defined models difficult to formulate and deceptive in their predictions.

For these reasons, this book is being published. Charge transfer in DNA is an area that will continue to intrigue, effectively because DNA systems are perhaps the limit of the complexity of charge transfer behavior. Most of the major progress in understanding charge transfer has come in beautifully simple and well-defined systems, ranging from binuclear metal complexes and well-defined intramolecular organic ET systems [7] to simple linkages or alkanes in molecular wire transport [102]. DNA is clearly much more complicated.

What does seem clear is that the kinds of motion that charges can exhibit in DNA (tunneling at short distances, hopping at long distances, vibronic control, polaron formation, coupling to ionic displacements, control by solvent polarization) are precisely those being explored for simpler ET systems. In this sense, DNA is not different, it is simply more complicated. Those complications remain fascinating, and as this entire book demonstrates, the DNA charge transfer problem is sufficiently rich and sufficiently intricate that no single set of models or experiments can clarify all of the behaviors. Perhaps the most fruitful path is to examine the most clearly defined systems, such as the hairpins whose photoexcited forward rates and relaxing backward rates have led to reproducible, interpretable, well-defined values for the rate constants and their energetic and temperature dependencies [103]. More measurements of this kind will continue to challenge the theoretical community to develop appropriate reduced models for understanding charge transfer and transport in well-defined DNA structures.

Acknowledgements We are grateful to the Chemistry Division of the ONR, MOLETRONICS program at DARPA, and to the DoD/MURI program for support of the research at Northwestern. The work at Duke is supported by NIH and NSF. We are grateful to many colleagues, particularly A. Troisi, J. Jortner, N. Rösch, D. Porath, and C. Dekker for sharing their insights with us.

References

1. Murphy CJ, Arkin MR, Jenkins Y, Ghatlia ND, Bossman SH, Turro NJ, Barton JK (1993) Science 262:1025
2. Arkin MR, Stemp EDA, Holmlin RE, Barton JK, Hormann A, Olson EJC, Barbara PF (1996) Science 273:475
3. Hall DB, Holmlin RE, Barton JK (1996) Nature 382:731
4. Jortner J, Bixon M, Langenbacher T, Michel-Beyerle ME (1999) Proc Natl Acad Sci USA 95:12759
5. Gutmann F (1967) Organic semiconductors. Wiley, New York
6. Jortner J, Ratner MA (1997) Molecular electronics. Blackwell Scientific, Oxford
7. Bixon M, Jortner J (1999) Adv Chem Phys 106:35
8. Olofsson J, Larsson S (2001) J Phys Chem B 105:10398
9. Newton MD (1991) Chem Rev 91:767
10. Schiff LI (1968) Quantum mechanics, 3rd edn. McGraw-Hill, New York
11. Lee M, Shephard MJ, Risser SM, Priyadarshy S, Paddon-Row MN, Beratan DN (2000) J Phys Chem A 104:7593
12. Tong GSM, Kurnikov IV, Beratan DN (2002) J Phys Chem B 106:2381
13. Risser SM, Beratan DN, Meade TJ (1993) J Am Chem Soc 115:2508
14. Priyadarshy S, Risser SM, Beratan DN (1996) J Phys Chem 100:17678
15. Beratan DN, Priyadarshy S, Risser SM (1997) Chem Biol 4:3
16. Priyadarshy S, Risser SM, Beratan DN (1998) JBIC 3:196
17. Brun AH, Harriman AJ (1994) J Am Chem Soc 116:10383
18. Meade TJ, Kayyem JF (1995) Angew Chem Int Ed 34:352
19. Jortner J, Bixon M, Voityuk AA, Rosch N (2002) J Phys Chem A 106:7599
20. Voityuk AA, Michel-Beyerle ME, Rösch N (2001) Chem Phys Lett 342:231
21. Voityuk AA, Jortner J, Bixon M, Rösch N (2000) Chem Phys Lett 324:430
22. Grozema FC, Siebbeles LDA, Berlin YA, Ratner MA (2002) Chem Phys Chem 3:536
23. Berlin YA, Burin AL, Siebbeles LDA, Ratner MA (2001) J Phys Chem A 105:5666
24. Grozema FC, Berlin YA, Siebbeles LDA (2000) J Am Chem Soc 122:10903
25. Grozema FC, Berlin YA, Siebbeles LDA (1999) Int J Quantum Chem 75:1009
26. Troisi A, Orlandi G (2001) Chem Phys Lett 344:509
27. Bixon M, Jortner J (2002) Chem Phys 281:393
28. Berlin YA, Burin AL, Ratner MA (2002) Chem Phys 275:61
29. Voityuk AA, Jortner J, Bixon M, Rösch N (2001) J Chem Phys 114:5614
30. Voityuk AA, Rösch N, Bixon M, Jortner J (2000) J Phys Chem B 104:9740
31. Voityuk AA, Rösch N (2002) J Phys Chem B 106:3013
32. Tavernier HL, Fayer MD (2000) J Phys Chem B 104:11541
33. Siriwong K, Voityuk AA, Newton MD, Rösch N (2002) J Phys Chem B 107:2595
34. Kurnikov IV, Tong GSM, Madrid M, Beratan DN (2002) J Phys Chem B 106:7
35. Eley DD, Spivey DI (1962) Trans Faraday Soc 58:411
36. Ladik J, Biczo G (1965) J Chem Phys 42:1658

37. Henderson PT, Jones D, Hampikian G, Kan YZ, Schuster GB (1999) Proc Natl Acad Sci USA 96:8353
38. Meggers E, Michel-Beyerle ME, Giese B (1998) J Am Chem Soc 120:12950
39. Giese B, Wessely S, Spormann M, Lindemann U, Meggers E, Michel-Beyerle ME (1999) Angew Chem Int Ed 38:996
40. Berlin YA, Burin AL, Ratner MA (2000) J Phys Chem A 104:443
41. Yu ZG, Song XY (2001) Phys Rev Lett 86:6018
42. Berlin YA, Burin AL, Ratner MA (2001) J Am Chem Soc 123:260
43. Giese B (2002) Curr Opin Chem Biol 6:612
44. Bixon M, Giese B, Wessely S, Langenbacher T, Michel-Beyerle ME, Jortner J (1999) Proc Natl Acad Sci USA 96:11713
45. Lewis FD, Li X, Liu J, Miller SE, Hayes RT, Wasielewski MR (2000) Nature 406:51
46. Lewis FD, Zuo X, Li J, Hayes RT, Wasielewski MR (2002) J Am Chem Soc 124:4568
47. Komineas S, Kalosakas G, Bishop AR (2002) Phys Rev E 65:061905
48. Basko M, Conwell EM (2002) Phys Rev Letters 88:098102
49. Landau LD, Lifshitz EM (1981) Physical kinetics. Pergamon, Oxford
50. Giese B, Amaudrut J, Köhler A-K, Spormann M, Wessely S (2001) Nature 412:318
51. Lewis FD, Liu JQ, Weigel W, Rettig W, Kurnikov IV, Beratan DN (2002) Proc Natl Acad Sci USA 99:12536
52. de Pablo PJ, Moreno-Herrero F, Colchero J, Gómez Herrero J, Herrero P, Baró AM, Ordejó P, Soler JM, Artacho E (2000) Phys Rev Lett 85:4992
53. Mirkin CA, Letsinger RL, Mucic RC, Storhoff JJ (1996) Nature 382:607
54. Loweth CJ, Caldwell WB, Peng XG, Alivisatos AP, Schultz PG (1999) Angew Chem Int Ed 38:1808
55. Taton TA, Mucic RC, Mirkin CA, Letsinger RL (2000) J Am Chem Soc 122:6305
56. Alivisatos P (2000) Pure Appl Chem 72:3
57. Demers LM, Mirkin CA, Mucic RC, Reynolds RA, Letsinger RL, Elghanian R, Viswanadham G (2000) Anal Chem 72:5535
58. Park SJ, Taton TA, Mirkin CA (2002) Science 295:1503
59. Cao YWC, Jin RC, Mirkin CA (2002) Science 297:1536
60. Nitzan A (2001) J Phys Chem A 105:2677
61. Berlin YA, Ratner MA (to be submitted)
62. Okahata Y, Kobayashi T, Nakayama H, Tanaka K (1998) Supramol Sci 5:317
63. Li X-Q, Yan Y (2001) Appl Phys Lett 79:2190
64. Cai L, Tabata H, Kawai T (2000) Appl Phys Lett 19:3105
65. Strom AJ, Van Noort J, De Vries S, Dekker C (2001) Appl Phys Lett 79:3881
66. Schuster GB (2000) Acc Chem Res 33:253
67. Rakhmanova SV, Conwell EM (2001) J Phys Chem B 105:2056
68. Kasumov AY, Kosiak M, Guéron S, Reulet B, Volkov VT, Klinov DV, Bouchiat H (2001) Science 291:280
69. Fink H-W, Schonenberger C (1999) Nature 398:407
70. Young MA, Ravishanker G, Beveridge DL (1997) Biophys J 73:2313
71. Barnett RN, Cleveland CL, Joy A, Landman U, Schuster GB (2001) Science 294:567
72. Porath D, Bezryadin A, De Vries S, Dekker C (2000) Nature 403:635
73. Warman JM, De Haas MP, Rupprecht A (1996) Chem Phys Lett 249:319
74. Tran P, Alavi B, Gruner G (2000) Phys Rev Lett 85:1564
75. Gómez-Navarro C, Moreno-Herrero F, de Pablo PJ, Colchero J, Gómez-Herrero J, Baró AM (2002) Proc Natl Acad Sci USA 99:8484
76. Bockrath M, Markovic N, Shepard A, Tinkham M, Gurevich L, Kouwenhoven LP, Wu MW, Sohn LL (2002) Nanoletters 2:187
77. Cai L, Tabata H, Kawai T (2001) Nanotechnology 12:211
78. Lee H-Y, Tanaka H, Otsuka Y, Yoo K-H, Lee J-O, Kawai Y (2002) Appl Phys Lett 80:1670
79. Yoo K-H, Ha DH, Lee J-O, Park JW, Kim J, Kim JJ, Lee H-Y, Kawai T, Choi HY (2001) Phys Rev Lett 87:198102-1

80. Watanabe H, Manabe C, Shigematsu T, Shimotani K, Shimizu M (2001) Appl Phys Lett 79:2462
81. Fink H-W (2001) Cell Mol Life Sci 58:1
82. Michael BD, O'Neill MP (2000) Science 287:1603
83. Landauer R (1957) IBM J Res Dev 1:223
84. Liang WJ, Shores MP, Bockrath M, Long JR, Park H (2002) Nature 417:725
85. Park J, Pasupathy AN, Goldsmith JI, Chang C, Yaish Y, Petta JR, Rinkoski M, Sethna JP, Abruña HD, McEuen PL, Ralph DC (2002) Nature 417:722
86. Klinger MI (1985) Usp Fiz Nauk 146:105
87. Buttiker M, Landauer R (1986) IBM J Res Dev 30:451
88. Nitzan A, Jortner J, Wilkie J, Burin AL, Ratner MA (2000) J Phys Chem B 104:5661
89. Yu ZG, Smith DL, Saxena A, Bishop AR (1999) Phys Rev B 59:16001
90. Burin AL, Berlin YA, Ratner MA (2001) J Phys Chem A 105:2652
91. Miller NE, Wander MC, Cave RJ (1999) J Phys Chem A 103:1084
92. Ponce A, Gray HB, Winkler JR (2000) J Am Chem Soc 122:8187
93. Brauns EB, Madaras ML, Coleman RS, Murphy CJ, Berg MA (1999) J Am Chem Soc 121:11644
94. Liang Z, Freed JH, Keyes RS, Bobst AM (2000) J Phys Chem B 104:5372
95. Toutounji MM, Ratner MA (2000) J Phys Chem A 104:8566
96. Hale PD, Pietro WJ, Ratner MA, Ellis DE, Marks TJ (1987) J Am Chem Soc 109:5943
97. Reimers J, Hush NS (unpublished)
98. Troisi A, Nitzan A, Ratner MA (2003) J Chem Phys 118:5356
99. Sugiyama H, Saito I (1996) J Am Chem Soc 118:7063
100. Su WP, Schrieffer JR, Heeger AJ (1980) Phys Rev B 22:2099
101. Bottger H (1985) Hopping conduction in solids. VCH, Deerfield Beach, Fla
102. Ratner MA, Reed MA (2002) Encyclopedia of Physical Science and Technology, 3rd edn, vol 10. Academic Press, New York
103. Lewis F (the chapter in this volume)

Top Curr Chem (2004) 237:37–72
DOI 10.1007/b94472

Quantum Chemical Calculation of Donor–Acceptor Coupling for Charge Transfer in DNA

Notker Rösch · Alexander A. Voityuk

Institut für Physikalische und Theoretische Chemie, Technische Universität München, 85747 Garching, Germany
E-mail: roesch@ch.tum.de
E-mail: voityuk@theochem.tu-muenchen.de

Abstract The electronic coupling V_{da} is the parameter which determines most strongly how the charge-transfer rate between donor and acceptor depends on the distance between the sites and the mutual orientation of donor and acceptor moieties. We discuss quantum chemical procedures to estimate electronic coupling matrix elements of hole transfer in DNA. The two-state model was shown to be quite reliable when applied to the coupling between neighboring Watson–Crick pairs. However, one has to be careful when employing the two-state model to estimate V_{da} in systems where donor and acceptor are separated by a bridge of base pairs. We considered the gross features of base-pair specificity, directional asymmetry, and conformation sensitivity of the couplings. Matrix elements between base pairs are found to be extremely sensitive to conformational changes of DNA. This strongly suggests that a combined QM/MD approach should be best suited for estimating V_{da} within DNA fragments.

Comparison of the effective couplings mediated by π-stack bridges TBT and ABA (B=A, ^{z}A, G, T, C) demonstrate that the efficiency of charge transfer is considerably affected by the nature of B; in turn, the effect of B strongly depends on the neighboring pairs. Especially large effects are due to the variation of the oxidation potential of guanine and adenine (B=G, A). Chemical modification of these species or changes of their environment strongly influence the efficiency of charge transfer.

We conclude with a discussion of several open questions and problems concerning the calculation of electronic couplings in DNA-related systems.

Keywords Electronic coupling · Charge transfer · DNA · Quantum chemical calculations

Abbreviations and Symbols

A, C, G, T	Nucleobases adenine, cytosine, guanine, and thymine, respectively. In DNA duplexes A, C, G, T stand for the corresponding Watson–Crick pairs, e.g., G in the duplex GGG corresponds to the (GC) Watson–Crick pair
^{z}A	7-Deazaadenine
AM1	Austin Model 1
AO	Atomic orbital
au	Atomic units
B3LYP	Hybrid Becke-3-parameter exchange and Lee-Yang–Parr correlation approximation

CNDO	Complete neglect of differential overlap
CSOV	Constrained space orbital variation (analysis)
CT	Charge transfer
DC	Divide-and-conquer (strategy)
DFT	Density functional theory
EA	Electron affinity
ET	Electron transfer
FC	Thermally weighted Franck–Condon factor
FCD	Fragment charge difference (method)
^{z}G	7-Deazaguanine
GMH	Generalized Mulliken–Hush (method)
HF	Hartree–Fock (method)
HOMO	Highest occupied molecular orbital
INDO	Intermediate neglect of differential overlap (method)
IP	Ionization potential
MD	Molecular dynamics
MNDO	Modified neglect of differential overlap (method)
MNDO/d	*MNDO* method, parameterization with d orbitals
MO	Molecular orbital
NDDO	Neglect of diatomic differential overlap (method)
NDDO-G	Special parameterization of the NDDO method
NDDO-HT	Parameterization of the NDDO method for hole transfer in DNA
PM3	Parameterized Model 3
QM/MD	Hybrid quantum mechanics/molecular dynamics (method)
SCF	Self-consistent field (method)
SFCD	Simplified fragment charge difference (method)
WCP	Watson–Crick pair
a	Acceptor
b	Bridge
d	Donor
k_{da}	Rate constant for charge transfer between donor and acceptor
V_{da}	Effective coupling between donor and acceptor states
H_{da}	Matrix element of Hamiltonian between diabatic donor and acceptor states
S_{da}	Overlap integrals between donor and acceptor states
Δ	Energy gap between adiabatic states
μ_1, μ_2	Dipole moments of the ground state and the first excited states, respectively
μ_{12}	Transition dipole moment
β, β_{el}	Decay parameter of the rate constant, decay parameter due to electronic contributions, respectively
λ, λ_i, λ_s	Reorganization energy, internal and solvent contributions, respectively

6-31G*	Gaussian basis set of so-called double-zeta quality for valence orbitals, augmented by polarization *d*-functions (*) on all atoms except hydrogen; used here to generate reference values of hole coupling matrix elements
6-311++G**	Very flexible Gaussian basis set of triple-zeta quality for valence orbitals, augmented by two sets of diffuse exponents (++) on all atoms (except hydrogen) and polarization functions on all atoms; other symbols for basis sets are to be read accordingly

1 Introduction

Electron transfer in extended systems including DNA, proteins, and molecular electronic devices continues to attract considerable interest [1–4]. Charge transfer (CT) in DNA is currently the subject of intense experimental [5–9] and theoretical [10–17] research; see also the contributions to this volume and references therein. It was shown experimentally that a guanine radical cation (G^+) can be generated in DNA far away from an oxidant because of the transport of a positive charge (hole transfer) [18, 19]. Based on guanines as resting states and guanine doublets or triplets as traps [7–9], two apparently contradictory mechanisms of charge migration in DNA were discussed and later on reconciled [20–24]: (1) single-step superexchange which is responsible for short-range CT separated by up to about 20 Å, and (2) multistep thermal hopping which goes far beyond that distance. In the first case, the CT process depends strongly on the nature and the length of the bridge between donor and acceptor [7–9]. In the latter case, the reaction rate depends only weakly on the distance between donor and acceptor. Such long-range hole migration in DNA can be treated as a series of superexchange steps between guanines separated by AT pairs [20–24]. Therefore, a microscopic study of CT within short DNA stacks is also important for a phenomenological description of long-range charge transport in DNA [14, 21, 23].

The transfer rate constant of single-step CT depends on various parameters [25, 26], but the electronic coupling V_{da}. is crucial for the dependence of the rate constant on the distance between a donor *d* and an acceptor *a* and on their orientation. Electronic interactions of donor and acceptor with the intervening medium, in turn, determine the coupling V_{da} which can be found from quantum chemical calculations on pertinent models. A number of excellent reviews discussed the quantum chemical treatment of electron transfer [27–29].

Thermal fluctuations are known to affect considerably the structure and other properties of biomolecules [30]. Recently, it was recognized that conformational changes in DNA can produce significant variations in the π-stacking of base pairs and thereby modulate the efficiency of charge transfer [31–33]. Thus, one has to employ a combination of molecular dynamics

(MD) and quantum-chemical calculations to obtain pertinent estimates of V_{da} on the relevant time scale.

This review focuses on computational schemes that can be applied to estimate the donor–acceptor electronic couplings in DNA. Therefore, we will ultimately be interested in a computational procedure that provides an efficient estimate of the electronic coupling when investigating a system along an MD trajectory [31]. In addition, sufficiently accurate methods and models play an important role in understanding fundamental aspects of the donor–acceptor coupling in DNA and in evaluating any procedure chosen for its efficiency in combination with an MD approach.

Unlike direct measurements of electrical conductivity of DNA [34, 35], chemical and photochemical experiments provide detailed data on how the CT efficiency depends on the DNA sequence and the local structure of an oligomer [5–9]. The latter experiments rely on intercalated or covalently bound chromophores which may affect the DNA structure. In the following, we will not discuss this effect of the chromophore although we realize that it may be important for a complete description of the systems used in those experiments. Rather, we will focus on a better understanding of the CT through unperturbed DNA fragments.

2 Methods

2.1 General Remarks

In a semiclassical picture, the rate k_{da} of nonadiabatic charge transfer between a donor d and an acceptor a is determined by the electronic coupling matrix element V_{da} and the thermally weighted Franck–Condon factor (FC) [25, 26]:

$$k_{da} = \frac{2\pi}{\hbar} |V_{da}|^2 (FC) \tag{1}$$

In line with the Franck–Condon principle, the electron transfer occurs at the seam of the crossing between diabatic (localized) states of donor and acceptor. The electronic coupling is the off-diagonal matrix element of the Hamiltonian defined at the crossing point.

From a fundamental point of view, one may prefer to determine the electron donor–acceptor coupling directly from diabatic states. This procedure has certain advantages [36], in particular when one is interested in a detailed investigation of electron correlation effects. Computational strategies that rely on adiabatic (delocalized) states are in general simpler to apply and thus more common [27].

Several such approaches for calculating electronic coupling matrix elements V_{da} have been proven to be useful. Most of them employ a two-state approximation [27, 28] where one assumes that donor and acceptor electronic states are well separated energetically from other states of the system; matrix elements have been determined with electronic wave functions obtained from either semiempirical or ab initio calculations [27, 28]. More general approaches beyond the two-state approximation are the block diagonalization procedure [37, 38], the generalized Mulliken-Hush method (GMH) [39, 40], and the fragment charge difference method (FCD) [41]. Thus far, the block diagonalization procedure has not been applied to DNA-related systems. The GMH and FCD schemes and their applications will be considered below in more detail.

2.2 Diabatic State Model

If the diabatic states corresponding to donor and acceptor are known, the electronic coupling V_{da} can be calculated as half of the energy gap Δ between the adiabatic states at the crossing of the diabatic states (H_{dd}=H_{aa}):

$$V_{da} = \Delta/2. \tag{2}$$

In the two-state model, the secular equation reads

$$\begin{vmatrix} H_{dd} - E & H_{da} - S_{da}E \\ H_{da} - S_{da}E & H_{aa} - E \end{vmatrix} = 0 \tag{3}$$

One directly obtains

$$V_{da} = \frac{\Delta}{2} = \frac{H_{da} - S_{da}\left(H_{dd} + H_{aa}\right)/2}{1 - S_{da}^2} \tag{4}$$

Usually the overlap S_{da} between donor and acceptor is small and therefore one has to first order:

$$V_{da} \approx H_{da} - S_{da}\left(H_{dd} + H_{aa}\right)/2 \tag{5}$$

In the rare case of orthogonal diabatic states, this reduces to:

$$V_{da} = H_{da}. \tag{6}$$

Thus, the electronic coupling is equal to the Hamiltonian matrix element H_{da} between donor and acceptor states. Recently, this relation was employed to estimate the coupling between nucleobases in DNA fragments [32]. Our estimates showed that both terms in Eq. 5, H_{da} and S_{da} (H_{dd}+H_{aa}), are of the same order of magnitude for the coupling between nucleobases. Thus, Eq. 6 is a rather crude and unnecessary approximation of Eq. 5.

2.3 Minimum Splitting Method

When donor and acceptor are equivalent by symmetry, then the electronic coupling can be estimated very easily as half of the energy gap Δ between the adiabatic states calculated at a reasonable geometry of the system [27]:

$$V_{da} = \frac{1}{2}(E_2 - E_1) \tag{7}$$

If donor and acceptor are "off resonance", an external perturbation should be applied to bring the *d* and *a* electronic levels into resonance or, equivalently, to minimize the adiabatic splitting. To this end, one can adjust the geometries of the relevant sites. However, this strategy is not very practical because, in general, one has to include also pertinent degrees of freedom of a polar environment [42], e.g., of a solvent which assists in the charge transfer. The effect of such fluctuations of a medium can be modeled in a very crude fashion by applying an external electric field, either induced by point charges [43–45] or directly as a homogeneous field [14, 46]. The electric field has to be suitably adjusted to bring the diabatic states of interest into resonance. This approach was applied to the coupling between isolated nucleobases [46] and Watson–Crick pairs (WCPs) in small DNA fragments [14]. Experience shows that the electronic coupling is not very sensitive to the strategy of how the resonance condition is achieved [27].

However, for systems with a weak *d–a* electronic interaction, $|V_{da}|<10^{-3}$ eV, minimization of the adiabatic gap to sufficient accuracy often is cumbersome; it becomes a real challenge when one has to treat an open-shell system. In any case, the minimum splitting method entails repeated quantum chemical calculations, a burden when very many couplings are to be evaluated along an MD trajectory. In that situation, one has to look for an alternative which permits direct treatment of off-resonance states.

2.4 Direct Treatment of Donor–Acceptor States

Within the two-state model, the *d–a* system can be described in terms of two adiabatic states ψ_1 and ψ_2 with energies E_1 and E_2. An orthogonal transformation

$$T = \begin{pmatrix} \cos\omega & -\sin\omega \\ \sin\omega & \cos\omega \end{pmatrix} \tag{8}$$

transforms ψ_1 and ψ_2 to the diabatic states $\tilde{\varphi}_d$ and $\tilde{\varphi}_a$, localized on donor and acceptor, respectively:

$$\tilde{\varphi}_d = \cos\omega\,\psi_1 + \sin\omega\,\psi_2$$

$$\tilde{\varphi}_a = -\sin\omega\,\psi_1 + \cos\omega\,\psi_2 \tag{9}$$

The wave functions $\tilde{\varphi}_d$ and $\tilde{\varphi}_a$ are orthonormalized and, therefore, $V_{da}=H_{da}$, see Eq. 6. The matrix element of the Hamiltonian $H_{da} = <\tilde{\varphi}_d|H|\tilde{\varphi}_a>$ can be expressed via the adiabatic energy gap:

$$H_{da} = \frac{1}{2}(E_2 - E_1)\sin 2\omega \tag{10}$$

Thus, to complete the procedure, one has to express the angle of rotation ω via the adiabatic states. Two such schemes, the generalized Mulliken–Hush method (GMH) [39, 40] and the fragment charge difference (FCD) method [41], have been suggested. The resulting coupling matrix elements will depend on the localization criterion used, but they are expected to be of similar magnitude if the physical nature of the localized donor and acceptor states is sufficiently well approximated by the transformation.

2.5 Generalized Mulliken–Hush Method

The GMH method of Cave and Newton [39, 40] is based on the assumption that the transition dipole moment between the diabatic donor and acceptor states vanishes, i.e., the off-diagonal element of the corresponding dipole moment matrix is zero. Thus, in the localization transformation one diagonalizes the dipole moment matrix of the adiabatic states ψ_1 and ψ_2. For a two-state model, the rotation angle ω can be expressed with the help of the transition dipole moment μ_{12} and the difference $\mu_1-\mu_2$ of the dipole moments of the ground state and the first excited state: $\tan 2\omega=2\mu_{12}/(\mu_1-\mu_2)$ [39, 40]. Then, according to Eq. 10, the electronic coupling is

$$H_{da} = \frac{(E_2 - E_1)\,|\mu_{12}|}{\sqrt{(\mu_1 - \mu_2)^2 + 4\mu_{12}^2}} \tag{11}$$

This expression reduces to Eq. 7 when donor and acceptor are in resonance, $\mu_1=\mu_2$. An advantage of Eq. 11 is that the coupling can also be estimated from experimental data [39, 40].

The GMH method is not restricted to two-state models; rather, it can be applied to systems with several pertinent electronic states. Very recently, a strategy was suggested of how to diagnose situations where a three-state treatment is advisable [47].

2.6 Fragment Charge Difference Method

We discuss the FCD method for hole transfer; a generalization to other cases of charge transfer is straightforward [41]. The (real) diabatic wave functions ϕ_d and ϕ_a are assumed to be normalized and completely localized on donor and acceptor, respectively. Furthermore, the adiabatic states ψ_1 and ψ_2 of the cationic system are assumed to include also a small contribution of a bridge b, e.g.:

$$\psi_1 = c_{d1}\varphi_d + c_{a1}\varphi_a + c_{b1}\varphi_b \tag{12}$$

The charge $q_i(f)$ localized on fragment f=d, a in the adiabatic state ψ_i can be measured by the value c_{fi}^2 if $|c_{bi}|<<1$ and $\langle\varphi_f|\varphi_b\rangle\approx 0$. In state $\tilde{\varphi}_d$ the charge on the donor is [41]:

$$\tilde{q}(d) = (c_{d1}\cos\omega + c_{d2}\sin\omega)^2 = q_1(d)\cos^2\omega + q_2(d)\sin^2\omega + q_{12}(d)\sin 2\omega \tag{13}$$

with $q_{12}(d)$=$c_{d1}c_{d2}$. A similar relation holds for the charge $\tilde{q}(a)$ on the acceptor in state $\tilde{\varphi}_a$. Then, the rotation angle ω is determined such that the charge transfer $\tilde{q}(d)+\tilde{q}(a)$ from donor to acceptor is maximized; in other words, in the new states $\tilde{\varphi}_d$ and $\tilde{\varphi}_a$, the charge is localized as much as possible on the donor and acceptor sites [41]. This leads to the condition:

$$\tan 2\omega = \frac{2\Delta q_{12}}{\Delta q_1 - \Delta q_2} \tag{14}$$

where Δq_{ij}=$q_{ij}(d)-q_{ij}(a)$ and Δq_{ii}=Δq_i. Combining Eqs. 10 and 14 we obtain:

$$H_{da} = \frac{(E_2 - E_1)\,|\Delta q_{12}|}{\sqrt{(\Delta q_1 - \Delta q_2)^2 + 4\Delta q_{12}^2}}, \tag{15}$$

The approach can also be applied to a system of N interacting donor, acceptor, or bridge sites, i.e., to a system with N fragments f_α (α=1,...,N) pertinent to CT. Then the set of diabatic states $\tilde{\varphi}_\alpha$ is defined such that the sum $\sum_\alpha \tilde{q}(\alpha)$ exhibits a maximum. The orthogonal transformation Θ from the N adiabatic states ψ_i to the diabatic states can be realized as a series of 2×2 transformations T, Eq. 8 [41]. Each such transformation T refers to one of the $\binom{N}{2}$ pairs of sites, using rotation angles defined by Eq. 14. Several iterations are usually needed until the maximum change of the charge $\tilde{q}(\alpha)$ on fragment α becomes sufficiently small, e.g., smaller than 10^{-5} e. Once the unitary transformation Θ is defined, the matrix elements of the diabatic Hamiltonian H can be calculated as H=$\Theta E\Theta^+$, where E is the diagonal matrix of the adiabatic energies E_i.

Finally, we mention a simplified FCD (SFCD) expression for the two-state model, where Eq. 15 is reduced to [41]

$$H_{da} = \frac{1}{2}(E_2 - E_1)\sqrt{1-\Delta q^2}. \tag{16}$$

Here, Δq is the difference of donor and acceptor charges in the ground state ψ_1. The SFCD model provides an estimate of the electronic coupling based only on the ground-state charge distribution and the vertical excitation energy; that information can be taken from experimental data. In the electronic ground state of a radical cation (or radical anion), fragment charges can be derived from EPR spectra via the spin density distribution. The SFCD expression, Eq. 16, can also be derived on the basis of the GMH approach [39, 40] or a related formalism [48, 49].

If donor and acceptor are in resonance, $\Delta q_1=\Delta q_2=0$ in Eq. 15 or $\Delta q=0$ in Eq. 16, then both FCD expressions reduce to Eq. 7. Also, when considering a CT system, it is often informative to estimate the energy gap $E_a - E_d$ between the diabatic donor and acceptor states. From Eq. 9, one directly obtains

$$E_a - E_d = H_{aa} - H_{dd} = (E_2 - E_1)\cos 2\omega = (E_2 - E_1)\,|\Delta q|. \qquad (17)$$

If donor and acceptor states are degenerate, $E_a=E_d$, then $\cos 2\omega=0$ and, therefore, $\sin 2\omega=1$; then, in line with Eq. 10, the electronic coupling is determined by half of the adiabatic energy splitting, Eq. 7.

2.7 Effective Hamiltonian Approach for DNA Fragments

The electronic coupling of donor and acceptor sites, connected via a π-stack, can either be treated by carrying out a calculation on the complete system or by employing a divide-and-conquer (DC) strategy. With the Hartree–Fock (HF) method or a method based on density functional theory (DFT), full treatment of a d–a system is feasible for relatively small systems. Whereas such calculations can be performed for models consisting of up to about ten WCPs, they are essentially inaccessible even for dimers when one attempts to combine them with MD simulations. Semiempirical quantum chemical methods require considerably less effort than HF or DFT methods; also, one can afford application to larger models. However, standard semiempirical methods, e.g., AM1 or PM3, considerably underestimate the electronic couplings between π-stacked donor and acceptor sites and, therefore, a special parameterization has to be invoked (see below).

In a DC strategy, one divides a system into overlapping fragments, calculates matrix elements for intra- and interfragment interactions, and then constructs an effective (one-electron) description for the whole system. For example, for hole transfer along a DNA oligomer d–$b_1b_2...b_n$–a, from hole donor d to acceptor a via a bridge of n WCPs b_i (i=1, 2,..., n), at least (n+2) states have to be treated, one per nucleobase pair: two states d^+–$b_1b_2...b_n$–a and d–$b_1b_2...b_n$–a^+ where the hole is localized on the donor and the acceptor, respectively, and n virtual states with the hole on one of the bridging subunits b_i. To estimate the diagonal elements $\widetilde{H}_{ii}$ of an effective Hamiltonian, one can resort to the energies of WCP trimer radical cations $b_{k-1}b_k^+b_{k+1}$ (or WCP dimers for the donor and the acceptor energies) [50]. In this way, significant effects of the environment of a localized hole state are accounted for [51]. As a first approximation, one estimates off-diagonal elements $\widetilde{H}_{ij} = V_{ij}$ of the effective Hamiltonian by the electron coupling between adjacent nucleobase pairs ($j=i\pm1$) [14] and neglects all other couplings. The state functions are assumed to be orthonormalized.

The effective Hamitonian approach [50] is analogous to schemes widely used for treating electron transfer in medium and large systems [27]. Of course, any model with one effective state per (bridge) unit represents an approximation because bridge states of higher energies can play a role for the effective donor–acceptor coupling.

Very recently, a thorough analysis of the DC method was carried out for several DNA fragments [16]. The calculated electronic couplings were found to be rather sensitive to the fragmentation scheme applied. The DC method converges more rapidly with fragment size for systems with larger energy gaps (at least 0.2 eV) between bridge states on the one hand, and donor and acceptor states on the other hand [16]. Therefore, if one aims at an accurate description of a system with a small tunneling energy gap between donor and bridge states, one has to model the electronic coupling by an effective Hamiltonian that is based on larger fragments [16].

2.8 Partitioning Scheme and Green Function Method

For systems where donor and acceptor are separated by a bridge, the effective coupling V_{da} can be estimated using Larsson's formula [52]

$$V_{da} = H_{d1} H_{na} \sum_{i=1}^{n} \frac{c_{1i} c_{ni}}{E - E_i}. \tag{18}$$

Here, H_{d1} is the coupling matrix element between the donor and the first site of the bridge; similarly, H_{na} is the matrix element between the last site of the bridge and the acceptor. The coefficients c_{1i} and c_{ni} characterize the contribution of the i-th bridge state (of energy E_i) at each end of the bridge. The tunneling energy E is an external parameter of the approach. When donor and acceptor are in resonance, $E{=}E_d{=}E_a$, E can be set to the average of E_d and E_a. Equation 18 was originally obtained with a partitioning technique [53]; later on, an alternative derivation was given with a Green function method [54, 55]. Priyadarshy et al. [56] used Larsson's formula to calculate the electronic coupling in DNA, taking into account the interference of distinct pathways within the bridge. We compared electronic couplings calculated with Larsson's approach [52] and the minimum splitting method using an effective Hamiltonian for DNA-related systems where donor and acceptor sites are separated by several intervening base pairs [50]. For bridges consisting of two or more WCPs, Larsson's scheme is clearly preferable because the effective coupling is calculated directly, whereas the second approach requires a very accurate iterative search for the minimum splitting, especially when V_{da} is 10^{-6} eV or less.

Equation 18 should be applied to estimate the donor–acceptor coupling H_{da} only if the virtual states of the bridge are not in resonance with donor or acceptor states. The electronic coupling between base pairs in a π-stack has to be notably smaller than the energy gap between the donor and bridge states. If a bridge state E_i would be below the tunneling energy E, then a bridge unit would trap the electron hole and the charge transfer would stop. Hence, this model should not be applied in cases where the energy splitting between bridge states and the donor becomes comparable to the thermal energy (0.026 eV at room temperature) because thermal injection of holes into a bridge would occur, i.e., the mechanism changes from superexchange to thermal hopping [10, 12, 13].

2.9 One-Electron Approximation

Different quantum chemical approaches can be invoked to calculate the electronic couplings. In many cases one can reliably estimate electron-transfer matrix elements on the basis of a one-electron approximation [27–29].

For ease of discussion, let us consider hole transfer between the nucleobase pairs guanine–cytosine (GC) and adenine–thymine (AT) in the 5′-3′ direction [14]. The radical cation states of the WCP doublets [(G^+C),(AT)] and [(GC),(A^+T)] are initial and final states of the hole transfer process. Recall that the purine nucleobases G and A have considerably smaller oxidation potentials than the pyrimidine bases C and T [57], and thus, in a simplified way of speaking, serve as donor and acceptor sites. The splitting $\Delta=E_2-E_1$ of the adiabatic electronic states is the first excitation energy of the radical cation [(G^+C),(AT)], which can be calculated using a configuration interaction method. Alternatively, invoking Koopmans' approximation, Δ can be estimated as the difference of the one-electron energies of the two highest occupied molecular orbitals HOMO and HOMO-1, calculated for the closed-shell neutral dimer [(GC),(AT)]. The latter procedure is much simpler, of course, and thus enjoys great popularity for calculating electron coupling matrix elements:

$$\Delta = E_2^{N-1} - E_1^{N-1} = \left(E_2^{N-1} - E_0^N\right) - \left(E_1^{N-1} - E_0^N\right) \approx \varepsilon_{HOMO-1}^N - \varepsilon_{HOMO}^N, \tag{19}$$

where the superscripts N and $N-1$ refer to the neutral and radical cation systems, respectively. $E_0{}^N$ is the ground-state energy of the neutral dimer [(GC),(AT)]; the energies ε_k with k=HOMO or HOMO-1 refer to the corresponding orbitals of that system. Δ values calculated with the HF–Koopmans approach and a complete active space method followed by a selected configuration interaction with the same basis set agree very well [58].

It is also practical to invoke a one-electron approximation in the FCD method [41] when one estimates donor and acceptor charges. Thus, one approximates the fragment charges of the radical cations [(G^+C),(AT)] and [(GC),(A^+T)] via the corresponding Mulliken populations of the HOMO and HOMO-1 of the neutral dimer. Then, the charge on fragment f in [(G^+C),(AT)] is

$$q_1(f) = \sum_{i \in f} c_{i,HOMO} \sum_{j=1}^{M} c_{j,\mathrm{HOMO}} S_{ij}. \tag{20}$$

Here, S_{ij} is the overlap of atomic orbitals i and j; i runs over atomic orbitals (AOs) associated with the selected fragment f while j runs over all AOs. The fragment charges of the second adiabatic state are calculated analogously using the coefficients $c_{i,\mathrm{HOMO\text{-}1}}$ of orbital HOMO-1 in place of $c_{i,\mathrm{HOMO}}$ in Eq. 20. By the same token, the quantity $q_{mn}(f)$ can be defined as

$$q_{mn}(f) = \frac{1}{2}\left[\sum_{i\in f} c_{i,HOMO+1-m} \sum_{j=1}^{M} c_{j,HOMO+1-n} S_{ij} + \sum_{i\in f} c_{i,HOMO+1-n} \sum_{j=1}^{M} c_{j,HOMO+1-m} S_{ij}\right]. \quad (21)$$

The matrix elements used in the FCD method, Eq. 15, are expressed as $\Delta q_{mn}=q_{mn}(d)-q_{mn}(a)$.

Similarly, to apply the GMH method, Eq. 11, one calculates the difference $\mu_1-\mu_2$ of the adiabatic dipole moments and the transition moment μ_{12} as follows:

$$\mu_1 - \mu_2 = \sum_{i,j=1}^{M} \left(c_{i,HOMO} c_{j,HOMO} - c_{i,HOMO-1} c_{j,HOMO-1}\right) D_{ij} \quad (22)$$

$$\mu_{12} = \sum_{i,j=1}^{M} c_{i,HOMO} c_{j,HOMO-1} D_{ij}. \quad (23)$$

Here, D_{ij} are the matrix elements of the dipole operator defined for AOs i and j.

Thus, the electronic couplings for hole transfer in cation radical systems can be treated using computational results for the corresponding neutral species. In this way, all parameters needed for the minimum splitting method as well as the FCD and GMH schemes can be efficiently estimated in a one-electron approximation, as given by the equations of this section.

3 Models

If one assumes that the two nucleobases of a WCP form a planar unit, then the mutual positions of two adjacent base pairs in DNA duplexes can be specified by six base step parameters (see Fig. 1): three translations—rise, shift, and slide; and three rotations—twist, tilt, and roll [59, 60]. In the ideal (reference) configuration, rise and twist are 3.38 Å and 36°, respectively, while all other step parameters are zero. When building models of double-stranded DNA fragment oligomers, we fully considered an accurate protocol, for instance as implemented in the program SCHNArP [60]. The structural parameters of the four bases adenine (A), cytosine (C), guanine (G), and thymine (T) can be derived from experimental data [61] or obtained from quantum chemical calculations. We found that small changes in the geometry of nucle-

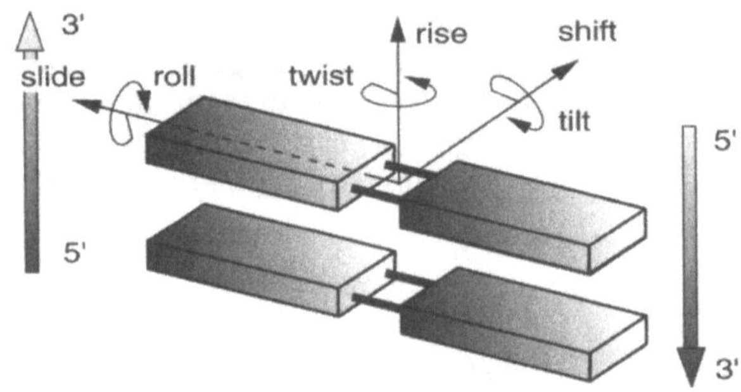

Fig. 1 The six step parameters that define the conformation of a Watson–Crick base pair in DNA

obases do not significantly affect the values of the couplings. Matrix elements calculated for experimental structures are very close to those obtained at B3LYP/6–31G* geometries; deviations amount to at most 5% [50].

In the DNA fragments considered below, base pairs are arranged in the conventional 5′→3′ direction [62]. We will consider only models of double-stranded DNA oligomers, comprising several stacked base pairs. The sugar-phosphate backbone was shown to play only a very minor role for charge transfer in DNA; its influence on electronic couplings between base pairs can be neglected [16, 56]. For instance, the effective electronic coupling between (GC) in the trimer [(GC),(AT),(GC)], calculated with and without the sugar-phosphate backbone, is 6.66×10^{-5} and 5.59×10^{-5} eV, respectively [16]. We do not include the sugar-phosphate backbone when calculating CT models.

Chemical modification of purine bases can induce considerable changes in the CT efficiency. The most important derivatives of purines, considered both experimentally and theoretically, are 7-deazaguanine (^{z}G) and 7-deaza-adenine (^{z}A) [63, 64]. Such modification may induce structural changes of the corresponding base pairs. Comparison of binding energies and structural parameters of hydrogen bonds in natural and modified pairs (GC) and (^{z}GC) and (AT) and (^{z}AT) calculated at the B3LYP/6–31G* level revealed only minor changes, both in the stability and the geometry of the pairs [50]. Based on these results, one does not expect substantial changes in DNA duplexes when the GC and AT pairs are replaced by ^{z}GC and ^{z}AT, respectively. Therefore, structural models of DNA fragments containing ^{z}G and ^{z}A can be designed using the same approach as for normal DNA.

4
Results of Quantum Chemical Calculations

Usually, we calculated the electronic couplings in DNA-related systems at the Hartree–Fock level [65]. For systems like [(GC),(CG)] and [(TA),(AT)], where donor and acceptor are equivalent by symmetry, the coupling matrix element can be estimated as half of the adiabatic energy splitting. Therefore, such symmetric systems cannot be used to compare the performance of different computational schemes. As indicated above, the GMH, FCD, and SFCD procedures simply reduce to Eq. 7 in such symmetric cases.

4.1
Effect of the Basis Set

In Table 1 we compare results for the electronic coupling between (GC) pairs in the dimer [(GC),(GC)], calculated with different basis sets. Although the mutual position of the pairs corresponds to ideal B-DNA, donor and acceptor are not symmetry-equivalent and are off resonance. With the standard basis set 6–31G*, the energy gap between both guanines is ~0.16 eV and their charges differ by ~0.7 e; for more extended basis sets, these values are even larger (Table 1). The matrix element decreases with increasing size of the basis set. Polarization and diffuse functions on hydrogen do not affect the result in an essential fashion; cf. the results for the basis sets 6–311+G* and 6–311++G** (Table 1). However, polarization functions on "heavy" atoms C, N, and O have a notable effect: the coupling calculated with the basis set 6–31G* is about 20% smaller than the value calculated with the basis set 6–31G. The effect of diffuse functions is relatively small, about 5%; cf. values calculated with the basis sets 6–31G* and 6–31+G*.

We chose the basis set 6–31G* as standard for HF calculations of the electronic couplings because it offers a reasonable compromise between accuracy and computational effort; see Table 1 for relative computer times.

Moreover, as will be shown below, electronic couplings are very sensitive to conformational changes of DNA fragments: their magnitude can change

Table 1 Effect of the basis set on the hole coupling matrix element V_{da} between the two guanine units of the WCP dimer [(GC),(GC)], calculated at the HF level with the GMH and FCD methods[a,b]

Basis set	E_2-E_1[c]	$\|E_d-E_a\|$[d]	Δq[e]	V_{da}		t_{rel}[f]
				GMH	FCD	
3–21G	0.256	0.161	0.628	0.0996	0.0997	0.17
6–31G	0.251	0.161	0.640	0.0965	0.0969	0.27
6–31G*	0.229	0.162	0.708	0.0807	0.0811	1
6–31+G*	0.224	0.166	0.741	0.0760	0.0760	12
6–311+G*	0.225	0.168	0.748	0.0757	0.0753	20
6–311++G**	0.224	0.168	0.752	0.0753	0.0747	40

[a] Energies in eV, charges Δq in e

[b] Adapted from [41]

[c] Adiabatic splitting estimated as energy difference of HOMO and HOMO-1 of the neutral system

[d] The donor–acceptor gap, from Eq. 17

[e] Difference of charges localized on the guanine moieties

[f] Computation time in relative units

by an order of magnitude or more, even for thermal structural fluctuations. Therefore, relative deviations of at most 10% seem quite acceptable in the present context. It is worth noting that GMH and FCD values of the electronic couplings, calculated within the same basis set, are very close to each other; deviations are about 2% at most (Table 1).

Also for other systems, V_{da} values for hole transfer depend only weakly on the basis set [14, 41, 46]. This is at variance with results for the coupling of excess electron transfer where energies of more diffuse states are involved; in the ground strategy this would correspond to employing energies of unoccupied molecular orbitals.

4.2 Semiempirical Methods

The first quantum chemical modeling of the effective electronic coupling in DNA was carried out with the semiempirical method CNDO/2 [56]. An important result of that study was that π-stacking interaction of nucleobases plays a dominant role. Most important was the conclusion that the experimentally found long-range charge migration in DNA, nowadays described as a hopping process [10, 12, 13], cannot be described with the superexchange mechanism.

Our calculations with several established semiempirical schemes (INDO/S [66], MNDO [67], AM1 [68], PM3 [69], and MNDO/d [70]) show that all these methods significantly underestimate the electronic coupling between π-stacked base pairs as compared with HF results. Typically, the matrix elements derived from semiempirical calculations are three to six times smaller (!) than the corresponding HF values.

We recently established such a semiempirical method on the basis of the NDDO-G scheme [71], which had been designed to simulate absorption spectra and ionization energies of large organic and biological molecules. With the same parameterization, the NDDO-G procedure predicts molecular geometries of organic molecules well at the SCF level; the mean absolute error of bond lengths is 0.014 Å and that of bond angles is 1.9°. The method provides electronic excitations with a mean absolute error of 0.13 eV; first and higher ionization potentials are reproduced with a mean absolute error of 0.24 eV. The NDDO-G method is suitable for studying structures of large organic and biological molecules and for interpreting and predicting their excited state properties. However, this semiempirical method cannot reproduce electronic couplings between stacked nucleobases with sufficient accuracy.

This limitation has been overcome with a special NDDO-HT parameterization for calculating hole coupling matrix elements in DNA-related systems [72]. As reference data, coupling matrix elements were calculated for a set of 130 structures of WCP dimers with different step parameters at the HF/6–31G* level. As discussed below in more detail, electronic couplings between neighboring pairs are extremely sensitive to conformational fluctuations of the DNA structure. For instance, the matrix element between base pairs in

[(AT),(AT)] ranges from 0.003 to 0.124 eV [31]. The new NDDO-HT parameterization allows one to reproduce these changes rather well. Relative deviations of the semiempirical estimates from the corresponding HF values are about 3% on average.

4.3 Estimating Electronic Couplings from Overlap Integrals

To a first approximation, the matrix element H_{da} is proportional to the overlap S_{da} for *d–a* distances of interest in DNA. Therefore, the coupling V_{da}, as given by Eq. 5, is approximately proportional to the overlap S_{da}. Troisi and Orlandi found an almost linear relationship between the electronic coupling of nucleobases and the overlap of the pertinent donor and acceptor orbitals. At the level HF/3–21G, the matrix element V_{da} (in eV) can be estimated as $-0.716\ S_{ij}$, where S_{ij} is the overlap integral calculated between the HOMOs of the donor and acceptor sites. This approximation obviously can be very useful when combined with MD simulations of DNA fragments. However, two remarks are in order: (i) the reference values of V_{da} should be generated with a more accurate method, e.g., based on Eq. 5 instead of Eq. 6 [32]; and (ii) the very small basis set 3–21G is insufficient for achieving satisfactory reference matrix elements (Table 1).

An overlap-based approach for estimating coupling matrix elements has a very important limitation. Unlike ab initio or semiempirical calculations, the scheme cannot be directly applied to systems where donor and acceptor sites are separated by an intervening bridge or medium. Thus, the scheme can be used only together with the effective Hamiltonian, constructed for models where only interactions between neighboring pairs are taken into account. Such models are well known to have limited accuracy in cases where only one state per bridge unit is taken into account [16]. Nevertheless, the suggested overlap-based approach seems to be useful for qualitative modeling of CT in DNA.

5 Electronic Couplings Between Neighboring Pairs

In this section we will consider some examples of the electronic coupling between neighboring Watson–Crick pairs, calculated in the two-state model.

5.1 Effect of the Donor–Acceptor Energy Gap

To study the effect of the energy gap between donor and acceptor, we carried out HF/6–31G* calculations on [(GC),(AT)] [41] and compared various procedures for determining the coupling elements within the one-electron approximation. In the complex [(GC),(AT)], guanine and adenine act as donor and acceptor sites. The donor–acceptor energy gap can be modulated by ap-

Table 2 Effect of an external electric field F on the coupling matrix element V_{da} between the G and A sites of the dimer [(GC),(AT)], calculated at the HF/6-31G* level with the GMH, SFCD, and FCD methods[a,b]

F	E_2-E_1	$\lvert E_d-E_a \rvert$	GMH			SFCD		FCD
			μ_{12}	$\mu_1-\mu_2$	V_{da}	Δq	V_{da}	V_{da}
8	1.557	1.509	−1.29	−15.67	0.1267	0.983	0.1439	0.1355
2	0.724	0.670	−2.81	−15.03	0.1269	0.933	0.1303	0.1307
0	0.467	0.387	−4.38	−13.48	0.1271	0.836	0.1281	0.1291
−2	0.275	0.102	−7.46	−6.04	0.1272	0.377	0.1272	0.1275
−2.715	0.255	0.001	−8.05	−0.01	0.1274	0.005	0.1273	0.1274
−4	0.315	0.183	6.52	9.45	0.1276	−0.578	0.1286	0.1281
−7	0.668	0.611	3.08	14.88	0.1280	−0.913	0.1365	0.1311
−13	1.501	1.455	1.38	15.73	0.1293	−0.976	0.1651	0.1378

[a] Electric field in 10^{-3} au, energies in eV, dipole moment matrix elements in Debye, charges Δq in e; for the definitions of various quantities, see Table 1
[b] Adapted from [41]

plying a weak external electric field perpendicular to the plane of the nucleobases. This perturbation also affects other characteristics relevant to charge transfer: fragment charges, the difference of the adiabatic dipole moments, and the transition dipole moment (Table 2).

As can be seen from Table 2, the applied electric field causes considerable changes of the *d–a* energy gap and the charge distribution. Without an external electric field, donor and acceptor in [(GC),(AT)] are rather far from resonance; the energy gap is 0.39 eV. The charges localized on guanine and adenine are 0.92 and 0.08 e, respectively; the dipole moment of the system changes by 13.5 Debye due to charge transfer. An electric field of 2.715×10^{-3} au brings donor and acceptor into resonance. As a result, their diabatic energies E_d and E_a differ by 0.001 eV and the charge is equally distributed on donor and acceptor (Table 2). In this case, the electronic coupling can be estimated as half of the adiabatic splitting $\Delta=E_2-E_1$, namely H_{da}=0.127 eV.

If the strength of the electric field increases, the donor–acceptor gap grows, the charge on guanine becomes progressively smaller (Δq becomes more negative), and the difference of the dipole moments $\mu_1-\mu_2$ increases (Table 2). However, the magnitude of H_{da} remains almost unchanged whereas the quantities entering the right-hand side of Eqs. 11, 15, and 16 change considerably. But even for very strong electric fields (e.g., 8×10^{-3} au or -13×10^{-3} au), where donor and acceptor are far from resonance, $|E_d-E_a|\approx1.5$ eV, the GMH and FCD schemes provide reliable estimates of the hole coupling matrix element H_{da} (Table 2).

In summary, these computational results suggest that the GMH and FCD methods are quite robust and can be applied to DNA fragments where donor and acceptor levels are far from resonance. On the other hand, the results demonstrate that the electron-transfer matrix element does not vary significantly when a perturbation by an external electric field is applied. This finding justifies the procedure for estimating the electronic coupling in DNA π-stacks where the splitting of adiabatic energies was minimized with the help of an external electric field [14, 46].

5.2 Electronic Couplings in Dimers

For understanding hole transfer in DNA, it is instructive to discuss the corresponding electronic matrix elements between base pairs in DNA of regular structure. In Table 3 we compare HF results for intrastrand and interstrand coupling, estimated for models comprising two WCPs, to results obtained for systems consisting of just two purine nucleobases [14]. Already the latter, simpler models permit one to analyze gross features of bridge specificity and directional asymmetry of hole-transfer electronic matrix elements between purine nucleobases in DNA duplexes.

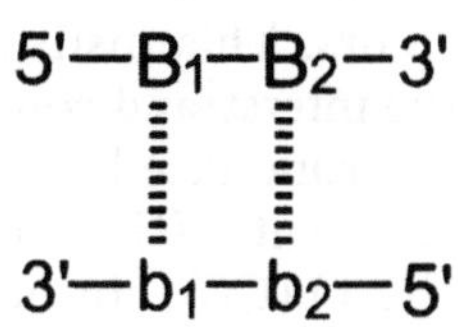

Scheme 1

By B_1–B_2, we denote intrastrand coupling between nucleobases B_1 and B_2, arranged along 5′-B_1–B_2-3′. Referring to bases of the opposite strand, complementary to B_i, as b_i (Scheme 1), we shall designate the two different con-

Table 3 Hole transfer coupling matrix elements (in eV) between two base pairs and comparison with results $V_{da}(B_1{:}B_2)$ of the corresponding two-base models[a,b]

Dimer of base pairs	V_{da}	Two-base model	$V_{da}(B_1{:}B_2)$
[(AT),(AT)], [(TA),(TA)]	0.026	A-A	0.030
[(AT),(TA)]	0.055	A\A	0.034
[(TA),(AT)]	0.050	A/A	0.062
[(GC),(AT)], [(TA),(CG)]	0.122	G-A	0.089
[(GC),(GC)], [(CG),(CG)]	0.093	G-G	0.084
[(GC),(CG)]	0.022	G\G	0.019
[(CG),(GC)]	0.078	G/G	0.043
[(AT),(GC)], [(CG),(TA)]	0.025	A-G	0.049
[(GC),(TA)], [(AT),(CG)]	0.026	G\A, A\G	0.021
[(TA),(GC)], [(CG),(AT)]	0.027	G/A, A/G	0.004

[a] Calculated at the HF/6-31G* level for models of regular structure. The orientation of the two purine bases B_1,B_2 is denoted as: B_1–B_2 for intrastrand configuration; B_1/B_2 and $B_1\backslash B_2$ for the two different interstrand configurations (see text)

[b] Adapted from [14]

figurations of interstrand pairs 5′-B_1...b_2-5′ and 3′-B_2...b_1-3′ as B_1/b_2 and $B_2\backslash b_1$, respectively. Here, slash and antislash are chosen to represent the structural schemes 5′-5′ and 3′-3′, respectively.

Each WCP dimer model consists of two purine bases (G and A) and two pyrimidine bases (C and T). According to the calculations, the two highest-lying orbitals HOMO and HOMO-1 of each duplex are mainly localized on the purine nucleobases, whereas the two occupied MOs following at lower energies, HOMO-2 and HOMO-3, are localized on pyrimidine nucleobases. Therefore, the purine–purine electronic coupling provides the dominant contribution to the hole transfer matrix elements, irrespective whether the bases belong to the same or to opposite strands.

The coupling depends crucially on the order of the bases; this can be traced to the specificity of the coupling between the purines. For instance, the couplings *V*(G-A) and *V*(A-G) were calculated as 0.122 and 0.025 eV, respectively. The electronic coupling between (GC) pairs with the guanines located on the same strand is significantly larger than the matrix elements for complexes, where the guanines are located on different strands. The opposite trend is found for (AT) pairs. This distinction can be rationalized in terms of the extraordinary large interstrand couplings A/A and A\A.

Hole transfer can exhibit a pronounced directional asymmetry. For instance, the coupling matrix element *V*(G-A) in [(GC),(AT)] is about five times larger than *V*(A-G) in [(AT),(GC)]. In both systems, G and A are in the same strand with the orientations 5′-G-A-3′ and 3′-A-G-5′, respectively. On the other hand, very similar electronic couplings G/A and G\A, with distinct interstrand orientation of G and A, are calculated for the systems [(GC),(TA)] and [(CG),(AT)], respectively.

Olofsson and Larsson determined electronic couplings in single- and double-stranded dimers, using HF/6–31G* and B3LYP/6–311G* calculations [17]. In most cases, HF and DFT results are in good agreement (within 15%); however, for several systems significant deviations were obtained. For instance, the intrastrand coupling between two adenine bases was calculated as 0.027 and 0.052 eV with HF and DFT methods, respectively [17]. A possible reason for this disagreement may be numerical inaccuracy of the minimum splitting method applied [17].

5.3
Effect of Pyrimidine Bases

Comparing matrix elements of simple models of two purine nucleobases with those calculated for WCP dimers (Table 3), one wonders about the effect of pyrimidine nucleobases on the electronic coupling matrix elements of hole transfer in DNA [14, 73].

Detailed analysis [74] of the pertinent molecular orbitals shows that it is impossible to identify any orbital-based direct involvement of the pyrimidine nucleobases in the hole transfer (see Fig. 2). Therefore, hole transfer always proceeds via purine bases, irrespective of whether they are arranged in the same or in opposite strands. However, the electronic coupling is notably

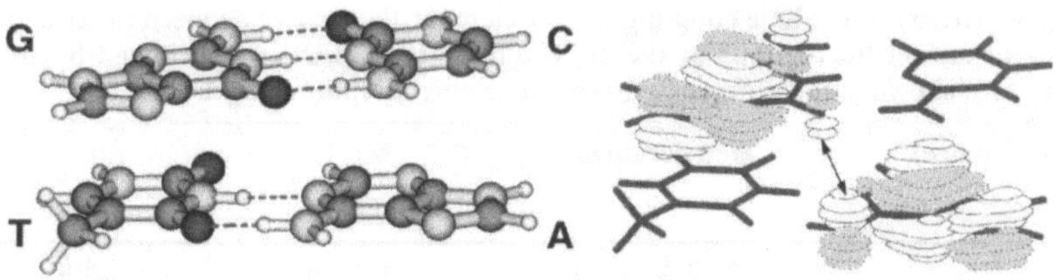

Fig. 2 Interstrand coupling for hole transfer illustrated for the HOMO-1 orbital of the dimer [(GC),(TA)]. *Arrows* indicate overlapping regions between purine orbitals

affected by the presence of pyrimidine bases, through electrostatic and exchange interactions between the nucleobases as well as by the formation of hydrogen bonds. A charge partitioning analysis [75] demonstrated that the polarization of purines due to the corresponding pyrimidine bases is remarkable, whereas the HOMO and HOMO-1 of the purine dimers remain mostly unchanged. A point-charge model confirmed the importance of the electrostatic interaction between purine and pyrimidine bases [74]. However, there are noteworthy deviations of the electronic coupling predicted by the point-charge model from the corresponding values obtained for dimers of the base pairs; they indicate an essential role of other interactions. A constrained space orbital variation (CSOV) analysis [76] revealed that the exchange interaction between purine and pyrimidine units plays a crucial role [74]; also, polarization of the pyrimidines leads to small corrections of the electronic coupling between the purine moieties (usually below 10%). Direct inspection of the spatial distribution of the HOMO and HOMO-1 of base-pair complexes provided a qualitative understanding of the electronic coupling, taking into account the linear relation between V_{da} and the corresponding orbital overlap (see Sect. 2.2). Also, hydrogen bonding between purine and pyrimidine bases clearly affects the electronic coupling.

Thus, when describing charge migration along the π-stack of DNA, it seems most appropriate to discuss the electron coupling between base pairs as the smallest units. In other words, less accurate results emerge if one corrects the direct electronic coupling of adjacent purine bases by perturbation theory, treating pyrimidines as bridge sites [46].

5.4 Effects of Structural Fluctuations

The electronic couplings in DNA π-stacks are very sensitive to conformational changes [31]. CT matrix elements were found to be very responsive to variations in the mutual positions of base pairs. As an example, let us compare the electronic couplings calculated for different conformations of the duplex [(GC),(GC)] (Table 4) [41]. In addition to the reference structure with a base-pair arrangement as in ideal B-DNA, we studied 12 distorted configurations; each of them differed from the reference by a single step pa-

Table 4 Sensitivity of the hole coupling matrix element V_{da} (in eV) between the two guanine units to structural fluctuations of the WCP dimer [(GC),(GC)] as calculated by the FCD scheme. Also shown is the difference Δq of donor and acceptor charges (in e)[a]

Ideal (Reference)	Step parameter	V_{da}	Δq
		0.0811	0.708
Rise	2.88 Å	0.1998	0.353
	3.88 Å	0.0294	0.916
Roll	5°	0.0482	0.888
	−5°	0.1256	0.461
Shift	0.5 Å	0.0198	0.952
	−0.5 Å	0.1364	0.562
Slide	1 Å	0.0033	0.992
	−1 Å	0.0556	0.816
Tilt	2°	0.0701	0.728
	−2°	0.0944	0.684
Twist	31°	0.0336	0.906
	41°	0.1044	0.649

[a] Adapted from [41]

rameter. The increments of the step parameters were chosen in line with their standard deviations extracted from X-ray data [59].

In all cases studied, the coupling strength estimated by the GMH and FCD schemes agree very well with the appropriate values calculated with the minimum splitting method [41]. Obviously, the electronic coupling between the guanine units of the dimer [(GC),(GC)] is very sensitive to conformational changes: its values range from 0.004 to 0.199 eV. Therefore, electron couplings between neighboring pairs in DNA can change considerably (by a factor of 50) due to thermal fluctuations of the helix conformation. One should keep in mind that the corresponding rate constant involves the square of the coupling matrix element (see Eq. 1).

As expected, the coupling in the dimer [(GC),(GC)] rises significantly as the distance between donor and acceptor decreases, from 0.029 eV at rise=3.88 Å to 0.199 eV at rise=2.88 Å (Table 4). The effect of other step parameters on the electron-transfer matrix elements can be rationalized with an analysis of the overlap between the HOMOs of donor and acceptor (see Fig. 2) [32, 74]. A shift of −0.5 Å leads to considerable increase of the coupling, whereas a shift of the same size but in opposite direction results in a remarkable decrease of the matrix element. Changing the angles between the base-pair planes may increase (roll=−5°, tilt=−2°, twist=41°) or decrease (roll=5°, tilt=2°, twist=31°) the matrix element. Thus, stabilization of special conformations of DNA fragments (e.g., with a negative shift parameter) can significantly enhance the CT efficiency. By the same token, one can expect that structural changes in chromophore–DNA complexes, which are widely used to study experimentally charge migration in DNA [5–9], have a significant direct structural effect on the CT kinetics.

5.5 Electronic Coupling within Watson–Crick Pairs

Electronic states of DNA fragments where the electron hole is located on a pyrimidine base are of high energy because cytosine and thymine are relatively weak hole acceptors [7, 51]. Nevertheless, the electronic coupling between nucleobases within (GC) and (AT) pairs may be of interest when one compares different CT paths in DNA using the superexchange model [14]. In Table 5, we present results of the electronic coupling *within* the base pairs (GC) and (AT) [41].

First, we note large energy gaps between donor and acceptor electronic states: 2.15 eV in (GC) and 1.50 eV in (AT). Accordingly, the charge is completely localized on the purine bases. FCD values are in good agreement with GMH results. In (GC), the GMH and FCD electronic couplings, calculated without external perturbation using the standard basis set, are 0.057 and 0.055 eV, respectively. The coupling between adenine and thymine is calculated somewhat smaller, 0.042 and 0.036 eV for GMH and FCD, respectively. With the very flexible basis set 6–311++G**, the coupling matrix elements change by 15–20% (Table 5). When one aims at an accurate estimate of the electronic coupling for CT between d and a lying in the same plane (e.g., hydrogen bonded d –a π-systems), the minimum splitting method employing a coplanar external electric field is not recommended because of the resulting strong artificial polarization of the π-system.

Table 5 Hole coupling matrix element V_{da} between purine and pyrimidine bases within (GC) and (AT) pairs calculated with the GMH and FCD methods[a,b]

Basis set		E_2–E_1	$\|E_d$–$E_a\|$	GMH			FCD	
				μ_{12}	μ_1–μ_2	H_{da}	Δq	H_{da}
GC	6–31G*	2.163	2.159	0.83	31.36	0.0569	0.998	0.0547
	6–311++G**	2.092	2.086	1.03	31.65	0.0679	0.997	0.0705
AT	6–31G*	1.505	1.502	0.88	31.34	0.0421	0.998	0.0363
	6–311++G**	1.462	1.459	1.02	31.62	0.0474	0.997	0.0425

[a] Energies in eV, dipole moment matrix elements in Debye, charges in au; for definition of various quantities, see Table 1

[b] Adapted from [41]

5.6 Systems with Three Donor–Acceptor Sites

Now we turn to the electronic coupling in the WCP trimers, as exemplified by the systems [(CG_1),(GC),(CG_2)] and [(T_1A_1),(GC),(T_2A_2)] in Scheme 2. For each trimer, we need to consider at least three relevant donor–acceptor centers: G, G_1, and G_2 in the first system; and G, A_1, and A_2 in the second system. The three HOMOs of each complex comprise the pertinent orbitals of these fragments. For the first complex, we consider the interstrand electronic couplings $V(GG_1)$ and $V(GG_2)$ and the next-nearest-neighbor intrastrand coupling between G_1 and G_2. The relevant matrix elements of the second complex are the interstrand couplings $V(GA_1)$, $V(GA_2)$, and the intrastrand coupling $V(A_1A_2)$.

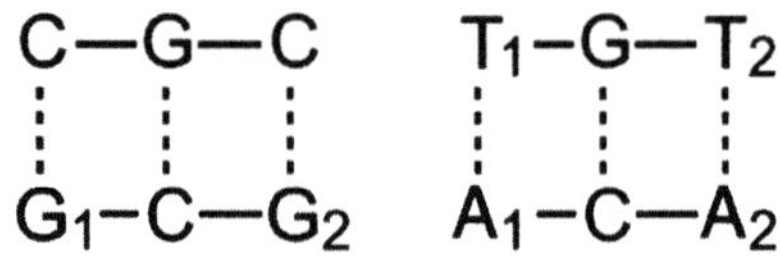

Scheme 2

First, comparing the GMH and FCD coupling values obtained within the three-state models [41], one can see that both computational procedures provide very similar results (Table 6). These values agree very well with electronic couplings calculated for suitable dimers using the minimum splitting method [14]. These results demonstrate that one can derive CT matrix elements between adjacent base pairs in DNA fragments from calculations of dimers, and that there is no essential effect of neighboring base pairs on the couplings.

However, in general, a multisite system cannot be reduced to a series of two-state systems. For instance, for calculating the couplings $V(G_1G_2)$ in the complex $[(CG_1),(GC),(CG_2)]$ and $V(A_1A_2)$ in $[(TA_1),(GC),(TA_2)]$, there does not seem to be an appropriate alternative to a simultaneous treatment of the three states. GMH and FCD calculations on $V(G_1G_2)$, using the two states with the largest contributions of the fragments G_1 and G_2, yield a coupling that overestimates the result of the three-state model by factor of 10, irrespective of the computational approach, GMH or FCD [41]. For the complex $[(TA_1),(GC),(TA_2)]$, the two-state approach underestimates the coupling

Table 6 Hole coupling matrix elements (in eV) in the WCP trimers $[(CG_1),(GC),(CG_2)]$ and $[(TA_1),(GC),(TA_2)]$ calculated with the GMH and FCD methods using three-state models and comparison with selected results of two-state models for WCP dimers[a]

Coupling	Three-state models		Two-state models[b]	
	GMH	FCD	Dimer	
G G_1	0.0779	0.0791	[(CG),(GC)]	0.078
G G_2	0.0217	0.0210	[(GC),(CG)]	0.022
G_1 G_2	0.0029	0.0029		
G A_1	0.0272	0.0270	[(TA),(GC)]	0.027
G A_2	0.0246	0.0240	[(GC),(TA)]	0.026
A_1 A_2	0.0045	0.0046		

[a] Adapted from [41]

[b] See Table 3

$V(A_1A_2)$; instead of 0.0045 eV (Table 6), the GMH and FCD methods both give about 0.0030 eV. Furthermore, in [(CG_1),(GC),(CG_2)], the energy of the bridge (GC) is very close to those of the donor and acceptor sites, (CG_1) and (CG_2). Therefore, one cannot estimate $V(G_1G_2)$ by a perturbation approach (superexchange model) or a related method [52, 54, 55]. Obviously, these systems cannot be reduced to a suitably coupled dimer when one considers coupling elements between nonadjacent sites.

Finally, for the example [(T_1A_1),(GC),(T_2A_2)], we will discuss the effect on the electronic couplings when the model is extended to more than three states, i.e., when one also takes into account states of higher energy that describe an electron hole localized on pyrimidine bases [41]. As fourth state, we select the state of next higher energy which is characterized by the HOMO-3 orbital, mainly localized on fragment T_1. Applying the FCD strategy to this four-state model, we calculated $V(GA_1)$=0.0257 eV, $V(GA_2)$= 0.0240 eV, and $V(A_1A_2)$=0.0046 eV. These values are very close to the couplings obtained within the three-state model (Table 6); differences amount to at most 5%. On the other hand, the corresponding GMH results of the four-state model are $V(GA_1)$=0.0189 eV, $V(GA_2)$=0.0246 eV, and $V(A_1A_2)$= 0.0038 eV. These values deviate considerably (up to 40%) from the data obtained for the three-state model (Table 6). If we add a fifth state, localized on fragment T_2 and described by a hole in HOMO-4, then the FCD coupling $V(A_1A_2)$=0.045 eV is very similar to that calculated with the three- and four-state models. However, the corresponding GMH value of 0.056 eV differs notably from the GMH results of the three- and four-state models.

As already noted by Cave and Newton [39], a transformation corresponding to a full diagonalization of the dipole moment matrix can be too restrictive in the following sense. If one adds one further state to a system described by N states, then N new constraints $\mu_{k,N+1}=0$ (k=1, 2,...,N) have to be accounted for. This can noticeably influence the resulting (approximate) diabatic states, hence the magnitude of the electronic coupling. Therefore, the number of states considered simultaneously within a GMH procedure should be kept as small as possible, to reflect only essential interactions in a system. The FCD method is more robust when applied to models with several moieties (or states per fragment) [41]. With the FCD transformation Θ (see Sect. 2.4), one obtains diabatic states that are localized on the various fragments. When one extends the number of fragments (or the states per fragment), states that are already localized are not affected in an essential fashion. Nevertheless, including states of arbitrary high energy may affect systems with small electronic couplings. Thus, for estimating CT matrix elements in systems comprising several donor–acceptor states it is recommended to take into account only low-lying adiabatic states relevant to charge transfer.

6 Effective Electronic Coupling in Duplexes with Separated Donor and Acceptor Sites

Recently, the modulation of DNA-mediated hole transfer for different base sequences was study experimentally, using 29-mer duplexes 5′-AGTGT GGG TTBTT GGG-3′ where two G triplets are separated by a bridge TTBTT [77]. B was one of the following purine nucleobases: adenine (A), 7-deazaadenine (^{z}A), guanine (G), and 7-deazaguanine (^{z}G). The nature of B considerably affects the efficiency of hole transfer between the two triple G sites; this effect was rationalized by differences of the ionization energy of B [77]. However, the ionization energy (or the oxidative potential) is not the only factor controlling the hole transfer efficiency. Chemical modification of the bases can also produce remarkable changes in the electronic couplings between Watson–Crick pairs. Yet, these two factors cannot be easily separated by experiments. Thus, motivated by these experimental findings [77], we carried out a theoretical study on the electronic coupling in systems containing 7-deazapurine bases [50] and we analyzed the effects of the bridge base B on the efficiency of DNA-mediated hole transfer.

Two main factors control how bridging bases affect the charge migration through the π-stack: (i) the energy of the virtual state B^+ and (ii) the electronic coupling of B (B=A, ^{z}A, G, ^{z}G) to adjacent nucleobases. As mentioned above, the relative energies of radical cation states B^+ depend in an essential fashion on adjacent nucleobase pairs [50]. To take this effect into account, the ionization energy of B was estimated in the trimer duplexes 5′-TBT-3′, 5′-ABA-3′, 5′-TBA-3′, and 5′-ABT-3′ with B=A, ^{z}A, G, ^{z}G and the electronic coupling matrix elements were calculated for the dimer duplexes 5′-TB-3′, 5′-BT-3′, 5′-AB-3′, and 5′-BA-3′. Our calculations showed that the chemical modifications G→^{z}G and A→^{z}A did indeed decrease the matrix elements by as much as 30% [50].

To estimate the consequences of such a bridge modification in a quantitative fashion one needs to treat an extended system where donor and acceptor sites are separated by intervening WCPs. We used the effective Hamiltonian approach described in Sect. 2.7 to clarify the distance dependence of the electronic coupling in DNA and the effect of bridge modifications [50].

6.1 Distance Dependence of Electronic Couplings

Consider two DNA fragments d–$b_1b_2...b_m$–a and d–$b_1b_2...b_n$–a where donor and acceptor are separated by π-stacked bridges of lengths m and n, respectively. The distance dependence of the rate constant is often expressed in terms of an exponential decay parameter β [5–9]:

$\ln k = -\beta(m+1)\Delta R + \text{const.}$, where ΔR=3.38 Å is the separation of adjacent pairs. For charge shift processes (as opposed to charge separation) between

Table 7 Effective electronic donor–acceptor coupling (in eV) mediated by various π-stacks of nucleobase pairs obtained with an effective Hamiltonian[a]

B	G	A	^{z}A	T	C
TBT	2.31×10^{-6}	4.21×10^{-6}	9.45×10^{-6}	1.68×10^{-6}	1.76×10^{-6}
TTBTT	4.66×10^{-9}	8.62×10^{-9}	19.4×10^{-8}	3.58×10^{-9}	3.86×10^{-9}
ABA	5.43×10^{-5}	4.80×10^{-6}	3.39×10^{-4}	1.16×10^{-5}	6.25×10^{-6}
AABAA	1.03×10^{-7}	1.02×10^{-8}	7.03×10^{-7}	2.46×10^{-8}	1.33×10^{-8}

[a] Adapted from [50]

identical moieties as donor and acceptor, e.g., guanine nucleobases, the free energy of the charge transfer is independent of the donor–acceptor separation. Then the distance dependence of the rate constant k, Eq. 1, is essentially due to $(V_{da})^2$ and the resulting decay parameter is only due to electronic factors, β_{el} (in Å^{-1}) which can be estimated as

$$\beta_{el} = -\frac{2}{3.38}\frac{\ln V_{da}(n) - \ln V_{da}(m)}{n-m}. \tag{24}$$

We applied the effective Hamiltonian approach to DNA fragments of standard geometry with $d=a=$GGG, determined the effective coupling from Eq. 18, and analyzed the coupling following Eq. 24. Now, we will discuss the results for various bridge compositions (Table 7) [51].

6.1.1 $(T)_n$, $(A)_n$, and $(AT)_{n/2}$ Bridges

The resulting values of the decay parameter β_{el} for bridges $(T)_n$ with n=1–6 are rather stable for various values of m and n in Eq. 24. The average value β_{el}=0.79 Å^{-1} agrees very well with a recent experimental value β=0.77 Å^{-1}, found for several systems where hole donor and acceptor are separated by AT pairs [78]. This value for a hole shift is intermediate between the values obtained previously for charge separation (β=0.7 Å^{-1}) and charge recombination (β=0.90 Å^{-1}) [79]. For bridges $(A)_n$ we calculated the same value as for bridges $(T)_n$: β_{el}=0.79 Å^{-1}. These results completely agree with experiments that yielded very similar values of β for DNA hairpins with guanines as hole acceptor, embedded either in poly-T or poly-A strands [79]. A study on strand cleavage in systems where single-step hole transfer occurs across $(AT)_n$ bridges yielded a value of β=0.7 Å^{-1} [8]. Comparing V_{da} for bridges $(AT)_n$ with n=2, 3, we calculate a very similar value, β_{el}=0.68 Å^{-1} [50]. Thus, despite identical parameters β_{el}=0.79 Å^{-1} calculated for $(A)_n$ and $(T)_n$ stacks, the decay parameter of an $(AT)_n$ stack differs notably. The difference is due to the fact that the intrastrand adenine–adenine interaction determines the coupling of intervening pairs in homogeneous bridges, whereas the interstrand A–A interaction is responsible for the coupling in $(AT)_n$ bridges (see Table 3).

6.1.2
TBT and ABA Bridges

Adenine and 7-deazaadenine exhibit similar value nearest-neighbor couplings H_{ij}, whereas their relative ionization energies differ significantly, calculated at 0.4 eV [50]. Therefore, as discussed above, a chemical modification of adenine makes it possible to elucidate experimentally the effect of the ionization energy of bridging nucleobases on the CT efficiency. Experiments [77] clearly showed that the efficiency of the hole transfer between two guanine triplets GGG, separated by a bridge TTBTT, is considerably affected by the nature of base B. In particular, the hole-transfer rate constant for B=^{z}A is substantially larger than that of the extremely inefficient case B=A.

Now we are in a position to illustrate with a computational model how the hole transfer capability of π-stack TBT bridging two triplets GGG changes when B=^{z}A replaces B=A (Table 7). *Inter*strand adenine–adenine coupling matrix elements determine the CT properties of the TBT bridge in both cases. The effective donor–acceptor coupling mediated by TAT and T^{z}AT were calculated as 4.2×10^{-6} and 9.5×10^{-6} eV, respectively [50]. This remarkable increase of the effective coupling is indeed due to the lower ionization potential of ^{z}A because the change of the base-pair couplings acts in the opposite direction: H_{da}(AT) and H_{da}(^{z}AT) were determined to be 0.55 and 0.43 eV, respectively [50]. While the values of the effective couplings decrease exponentially with the length of the stack when bridges are extended from TBT to TTBTT, the *ratio* of the effective couplings remains unchanged in our model. This result follows directly from Eq. 18 if one represents the system *d*–TTBTT–*a* as *d'*–TBT–*a'* with *d'*=*d*–T and *a'*=T–*a*. For instance, the couplings calculated for TTATT and TT^{z}ATT are notably smaller, 8.6×10^{-9} and 19.4×10^{-9} eV, respectively, than those of TAT and T^{z}AT given above. However, their ratio, about 2.3, remains unchanged. Thus, the effective coupling of $(T)_nA(T)_n$ stacks increases by a factor of about 2.3 when adenine is replaced by 7-deazaadenine. This result is in agreement with experiments [77], which demonstrated that the charge transfer mediated by the bridge TTATT becomes considerably more efficient on the substitution A→^{z}A. Unfortunately, these experimental results were not quantified, preventing a more detailed comparison.

A much more pronounced effect was found for π-stacks ABA where the *intra*strand purine–purine interaction is responsible for the electronic coupling. The effective donor–acceptor couplings for AAA and A^{z}AA were calculated as 4.8×10^{-6} and 3.4×10^{-4} eV, respectively (Table 7). The considerable increase of the effective coupling V_{da} by a factor of ~70 is due to the small difference of the ionization potentials of A^{z}A^{+}A and GG^{+}G, 0.019 eV [50]. Note that this small energy difference puts these systems at the limit of the Green function approach [52].

Furthermore, we compare the effective electronic couplings for stacks TBT, TTBTT, ABA, and AABAA where B=A, ^{z}A, G, T, and C (Table 7) [50]. As expected, the effects of B on the couplings provided by three- and five-membered bridges are similar. Despite favorable energetics, the effective coupling mediated by the sequence A-A-A in the system *d*–TTT–*a* is relatively small, because of the weak interstrand donor–bridge and bridge–acceptor interaction as well as the small intrastrand A-A coupling. In the bridge TAT, the effective coupling occurs via interstrand interactions between adenines that were calculated twice larger than the intrastrand coupling A-A. Therefore, the bridge TAT couples donor and acceptor stronger than the bridge TTT, despite the less favorable hole transfer energetics of TAT as compared to AAA [50]. Note that when comparing the bridges TTT and TAT one implicitly compares two "paths", *d*\A-A-A/*a* and *d*\A/A\A/*a*.

For different nucleobases B, the effective donor–acceptor coupling mediated by the stacks TBT increases in the order $\mathrm{T} \approx \mathrm{C} < \mathrm{G} < \mathrm{A} < {}^{z}\mathrm{A}$, whereas the coupling provided by ABA bridges increases in the order $\mathrm{A} < \mathrm{C} < \mathrm{T} < \mathrm{G} < {}^{z}\mathrm{A}$ (Table 7) [50]. Thus, the effect of a nucleobase B in the bridge XBY depends considerably on the neighboring WCPs X and Y; this is mainly due to the remarkable variation of the electronic couplings between B and its neighbors. Furthermore, our results for both types of bridge, TBT and ABA, suggest that the hole transfer efficiency increases notably if one replaces A by ^{z}A. This finding, completely in line with the experiments of Saito et al. [77], is a consequence of two facts: (i) the ionization potential of ^{z}A is considerably lower than that of A, and (ii) adenine and 7-deazaadenine exhibit comparable electronic coupling values to neighboring nucleobases.

We close with a discrepancy between experimental and computed results. The effective couplings in TTBTT do not agree with the normalized intensities of cleavage bands [77]. According to experiment, the observed efficiency changes with B as $\mathrm{A} < {}^{z}\mathrm{A} \approx \mathrm{G}$ whereas the calculated values of the electronic coupling vary in the order $\mathrm{G} < \mathrm{A} < {}^{z}\mathrm{A}$. This disagreement may be related to a *different* (and notable) variation of the electronic couplings with the structure of the duplex; see Sect. 5.4 [31, 41].

7 Open Questions

Finally, we turn to a series of topics which are important for a detailed understanding of charge transport in DNA, but so far have not been treated at a sufficiently comprehensive level.

7.1 Quantum Chemical Treatment of Electronic Couplings in DNA Fragments

In our review, we distinguished two types of DNA-related donor–acceptor systems: models where donor and acceptor interact directly, and models where they are separated by intervening base pairs.

If one attempts an accurate description of the coupling between neighboring pairs, one needs to answer several general questions. How accurate is the one-electron approximation? How essential are the effects of electron correlation on the coupling matrix elements? Whereas these points have already been considered for several simple CT systems, so far such calculations for DNA-related systems are still lacking.

In addition, several important questions arise when one discusses large DNA fragments consisting of several base pairs. Even cases where the whole system of interest can be treated by a sufficiently accurate quantum chemical method (often this is impossible), estimates of the electronic coupling within the two-state model may lead to inaccurate results, as demonstrated for trimer duplexes [41].

Obviously, the situation will be more difficult when additional approximations have to be employed, e.g., the divide-and-conquer scheme, an effective Hamiltonian or a perturbation approach. A central open question is which electronic states of the fragments have to be included so that reliable results, as compared with supermolecular calculations, can be provided. In any case, accounting for just one state per base pair will yield only semiquantitative results. A careful analysis of this point is highly desirable for DNA-related systems.

7.2 Effect of the Reorganization Energy on the Coupling

As already noted, the effective donor–acceptor electronic coupling depends considerably on the gap between donor and bridge levels or, as sometimes referred to, on the tunneling energy. Marcus and Sutin pointed out [42] that the reorganization energy λ has to be taken into account when one estimates the energy gap; the corresponding correction of the gap is $\lambda/2$. Commonly one distinguishes two contributions to the reorganization energy and both have been discussed for DNA-related systems [15–17, 80, 81]: the internal reorganization λ_i and the solvent contribution λ_s. λ_i is associated with geometry changes of donor and acceptor and can be estimated from both experimental data (as difference of vertical and adiabatic ionization potentials of the corresponding species) and quantum chemical calculations. At the B3LYP/6–31G* level of calculation, the internal reorganization energy for charge transfer between (GC) pairs was calculated as ~17 kcal/mol [81].

The solvent reorganization energy increases with the distance between donor and acceptor. Various theoretical estimates of λ_s for hole transfer in similar DNA fragments differ significantly [16, 80, 81], depending on the computational strategy. Calculated values of λ_s for systems with a *d–a* separation of ~10 Å (two intervening nucleobase pairs) range from ~31 kcal/mol [16] to ~46 kcal/mol [81] and ~69 kcal/mol [80]. Delocalization of the hole over two bases (see below) decreases the solvent reorganization by ~12 kcal/mol [81]. On the other hand, estimates inferred from experiment [9, 82] are notably smaller than the *calculated* values, ~10–40 kcal/mol for 0–3 intervening pairs.

A λ value of ~2 eV would have a considerable effect on the effective electronic coupling in DNA because it would imply a substantial increase of the gap between donor and bridge, and thus would entail a significant decrease of the coupling, e.g., by a factor of 40 between the guanines in the duplex GAAAG [16]. Moreover, the calculated exponential decay parameter β_{el} of the CT rate would increase dramatically [16] and become much larger than measured values. This leads to the conclusion that the superexchange mechanism has to be ruled out for bridges comprising not more than two intervening base pairs between donor and acceptor sites [16].

7.3 Delocalization of Hole States in DNA

Two complementary theoretical studies on hole localization in DNA were recently performed [15, 17]. Kurnikov et al. showed that two counteracting effects determine the hole charge distribution in DNA fragments. Whereas delocalization of the charge reduces the intrinsic energy of the system, the interaction with a polar environment stabilizes states with a localized hole (see below) [15]. The size of the hole in DNA was estimated to range from one to three base pairs [15].

Olofsson and Larsson showed [17] that in some cases of periodic DNA, such as C_n,T_n, and $(AT)_n$, hole states should be delocalized. According to their criterion, charge localization should occur if the internal reorganization energy exceeds the coupling matrix element by a factor of 4. However, the solvent reorganization energy contribution for CT in DNA seems to be larger than the internal reorganization energy (see above) [80, 81] and, therefore, localized states should be favored. In summary, one can expect hole states in DNA to be localized to a large extent [15, 17]. However, more sophisticated modeling, accounting for dynamical features of DNA and its environment, may lead to a different conclusion (see below).

7.4 Proton Transfer Coupled to Electron or Hole Transfer

Proton transfer within base pairs, and between bases and their environment, can play an important role for electron transfer processes in DNA [62]. Because electron hole trapping by a nucleobase can entail significant changes of the proton affinities of the relevant centers in a nucleobase, one can expect that proton transfer couples to electron transfer. Steenken suggested that proton transfer can interrupt the hole transfer process along a DNA π-stack [83]. Conversion of radical cations to neutral radicals due to proton transfer can create a CT driving force, and can decrease the effective electronic coupling mediated by the bridge due to an enlarged energy gap between donor and bridge states. In addition, proton transfer will change the direct coupling between nucleobases involved in the proton transfer. A very recent theoretical study [84] revealed that proton-coupled electron transfer in a thymine–acrylamide complex is highly sensitive to the solvation properties of the system and, therefore, a very accurate treatment of environmental effects is required.

7.5 QM/MD-Based Estimates of Electronic Couplings

Recall that the electronic coupling between base pairs in DNA is very sensitive to conformational changes and hence it varies considerably with time [31, 41]. In our first study of this topic [31], we considered several conformations of the dimers [(TA),(TA)], [(AT),(TA)], and [(TA),(AT)] embedded in a DNA duplex. We found the intrastrand A-A interaction to be more susceptible to conformational changes than the corresponding interstrand interaction. Our analysis of the effective electronic coupling between guanines in the duplex CCAACGTTGG, carried out on structures extracted from an MD trajectory, demonstrated that the rate of charge migration as measured by the square of the electronic coupling matrix element can vary several hundred-fold in magnitude, due to moderate changes of the duplex conformation. Thus, thermal fluctuations of the DNA structure have to be taken into account when one aims at a realistic description of the electron hole transfer in DNA.

In that study [31], we estimated the electronic coupling with the help of HF/6–31G* calculations. Any attempt to expand such an investigation into a reasonably quantitative description of the variation of the electronic coupling over time would be much too costly. As noted above, one can overcome that problem by constructing a special semiempirical method (e.g., NDDO-HT) affording sufficiently accurate estimates of electronic matrix elements, or by using an approximate relation between H_{da} and the overlap of related orbitals.

The latter strategy was used to estimate electronic couplings between guanines separated by the bridges AAAA, TATA, and TTTT [32]. During a short period of time (100 fs), the value of the *d–a* effective coupling can increase by several orders of magnitude when all related matrix elements between neighboring bases are rather high by coincidence. The average CT rate was found to increase by two orders of magnitude due to thermal fluctuations of the DNA structure. The Fourier transform of the time-dependent coupling $V_{da}(t)$ showed one strong peak at zero frequency; this was interpreted as domination of stochastic processes due to interference between several oscillators that are frequently dephased by interaction with a thermal bath [32].

Recently, the effects of static and dynamic structural fluctuations on the electron hole mobility in DNA were studied using a time-dependent self-consistent field method [33]. The motion of holes was coupled to fluctuations of two step parameters of a duplex, rise and twist (Fig. 1), namely the distances and the dihedral angles between base pairs, respectively. The hole mobility in an ideally ordered poly(G)-poly(C) duplex was found to be decreased by two orders of magnitude due to twisting of base pairs and static energy disorder. A hole mobility of 0.1 $cm^2V^{-1}s^{-1}$ was predicted for a homogeneous system; the mobility of natural duplexes is expected to be much lower [33]. In this context, one can mention several theoretical studies, based on band structure approaches, to estimate the electrical conductivity of DNA [85–87].

7.6
Beyond the Semiclassical Picture

As already discussed, the electronic coupling of donor and acceptor is extremely sensitive to conformational changes of DNA. The matrix element squared varies by a factor of three orders of magnitude due to thermal fluctuations. Thus, the Condon approximation may fail for charge transfer in DNA. This problem is discussed in more detail in the contribution of Beratan et al. to this volume. Consequently, instead of using Eq. 1, one has numerically to integrate the trajectory of a system for ~1 ns to estimate the CT rate constant; along this trajectory one has to calculate the electronic coupling for each DNA conformation. Troisi and Orlandi demonstrated [32] that one can treat charge transfer in DNA at a new qualitative level by combining MD simulations of DNA fragments with approximate estimates of the electronic coupling. In this context, a general analysis [88] may be of interest, which shows how vibronic coupling and anharmonicity of vibrational modes of a CT system affect the nonadiabatic charge-transfer rate.

7.7
Excess Electron Transfer

The majority of experiments on charge transfer in DNA deal with hole transfer. There are only few experimental studies pertaining to the transfer of excess electrons in DNA [89–91]. Theoretical studies focused on electron affinities of bases and their complexes [92–95].

When estimating the energetics of excess electron transfer in DNA via differences of electron affinities (EA) of nucleobases B in WCP trimers 5′-XBY-3′ [92], we found the EA values of bases to decrease in the order C≈T>>A>G. The destabilizing effect of the subsequent base Y is more pronounced than that of the preceding base X. As strongest electron traps, we predicted the sequences 5′-XCY-3′ and 5′-XTY-3′, where X and Y are pyrimidines C and T. These triads exhibit very similar EA values, and therefore, the corresponding anion radical states should be approximately in resonance, favoring efficient transport of excess electrons in DNA [92].

Electronic coupling matrix elements for DNA-related systems are much more difficult to calculate than the coupling of hole transfer. First, in the case of an excess electron, the one-electron approximation likely is insufficient and electron correlation is expected to play a crucial role. Second, preliminary results revealed a considerable influence of the basis set on the calculated coupling. Third, an excess electron is expected to be delocalized over several pyrimidine bases; this will render the evaluation of V_{da} even more difficult. Thus far, no reliable estimates of electronic coupling matrix elements seem to be available.

7.8
Concluding Remarks

Despite considerable effort and remarkable progress in quantum chemical calculations of electronic couplings V_{da} for hole transfer in DNA, it is still a challenge to estimate this key parameter in a reliable fashion by computational modeling. We pointed out a number of open questions: limitations of the two-state model and of perturbation approaches, delocalization of the charge over several nucleobase pairs, and the influence of the reorganization energy on V_{da}. In particular, the electronic couplings exhibit an unexpectedly strong sensitivity to thermal fluctuations of the DNA structure and its environment. This limits the validity of the semiclassiscal nonadiabatic description, Eq. 1, of the charge transfer in DNA.

In addition, models will need to change when an intercalated or bound chromophore is taken into account. Two issues should be considered in this context: structural changes of the DNA fragment due to its interaction with chromophore and photochemical hole injection into the π-stack.

Clearly, a detailed and comprehensive analysis of dynamical effects of the environment, in particular of fluctuations of the electrostatic potential in the interior of the DNA duplex created by the hydrated sodium ions of surrounding electrolyte, and the influence of these fluctuations on charge transfer parameters is desirable.

Acknowledgements We thank M. Bixon, J. Jortner, A. Marquez, M.E. Michel-Beyerle, M.D. Newton, J. Rak, and K. Siriwong for stimulating discussions and various contributions to the work described here. Our research was supported by Deutsche Forschungsgemeinschaft (SFB 377), Volkswagen Foundation, and Fonds der Chemischen Industrie.

References

1. Bixon M, Jortner J (eds) (1999) Electron transfer from isolated molecules to biomolecules. Adv Chem Phys, vols 106 and 107
2. Jortner J, Ratner M (eds) (1997) Molecular electronics. Blackwell, Oxford
3. Cornil J, Beljonne D, Calbert JP, Bredas JL (2001) Adv Mater 13:17
4. Niemeyer C (2001) Angew Chem Int Ed 40:4128
5. Hall DB, Holmin RE, Barton JK (1996) Nature 382:731
6. Nunez ME, Barton JK (2000) Curr Opin Chem Biol 4:199
7. Schuster GB (2000) Acc Chem Res 33:253
8. Giese B (2001) Acc Chem Res 33:631
9. Lewis FD, Letsinger RL, Wasielewski MR (2001) Acc Chem Res 34:159
10. Bixon M, Jortner J (2001) J Am Chem Soc 123:12556
11. Ratner MA (1999) Nature 397:480
12. Berlin YA, Burin AL, Ratner MA (2001) J Am Chem Soc 123:260
13. Berlin YA, Grozema FC, Siebbeles LDA (2000) J Am Chem Soc 122:10903
14. Voityuk AA, Jortner J, Bixon M, Rösch N (2001) J Chem Phys 114:5614
15. Kurnikov IV, Tong GSM, Madrid M, Beratan DN (2002) J Phys Chem B 106:7
16. Tong GSM, Kurnikov IV, Beratan DN (2002) J Phys Chem B 106:2381
17. Olofsson J, Larsson S (2001) J Phys Chem B 105:10398
18. Meggers E, Michel-Beyerle ME, Giese B (1998) J Am Chem Soc 120:12950

19. Giese B, Wessely S, Spormann M, Lindemann U, Meggers E, Michel-Beyerle ME (1999) Angew Chem Int Ed 38:996
20. Bixon M, Giese B, Wessely S, Langenbacher T, Michel-Beyerle ME, Jortner J (1999) Proc Natl Acad Sci USA 96:11713
21. Bixon M, Jortner J (1999) Adv Chem Phys 106:35
22. Jortner J, Bixon M, Langenbacher T, Michel-Beyerle ME (1998) Proc Natl Acad Sci USA 95:12759
23. Bixon M, Jortner J (2000) J Phys Chem B 104:3906
24. Berlin YA, Burin AL, Ratner M (2000) J Phys Chem A 104:443
25. Marcus RA (1964) Annu Rev Phys Chem 15:155
26. Marcus RA, Sutin N (1985) Biochem Biophys Acta 811:265
27. Newton MD (1991) Chem Rev 91:767
28. Newton MD (1999) Adv Chem Phys 106:303
29. Mirkin CA, Ratner MA (1992) Annu Rev Phys Chem 43:719
30. Cheatham TE, Kollman PA (2000) Annu Rev Phys Chem 51:435
31. Voityuk AA, Siriwong K, Rösch N (2001) Phys Chem Chem Phys 3:5421
32. Troisi A, Orlandi G (2002) J Phys Chem B 106:2093
33. Grozema FC, Siebbeles LDA, Berlin YA, Ratner MA (2002) Chem Phys Chem 3:536
34. Fink HW, Schönenberger C (1999) Nature 398:407
35. Porath D, Bezryadin A, de Vries S, Dekker C (2000) Nature 403:635
36. Sanz JF, Malrieu JP (1993) J Phys Chem 97:99
37. Pacher T, Cederbaum LS, Köppel H (1993) Adv Chem Phys 84:293
38. Domcke W, Woywood C, Stengle M (1994) Chem Phys Lett 226:257
39. Cave RJ, Newton MD (1997) J Chem Phys 106:9213
40. Cave RJ, Newton MD (1996) Chem Phys Lett 249:15
41. Voityuk AA, Rösch N (2002) J Chem Phys 117:5607
42. Marcus RA, Sutin N (1985) Biochim Biophys Acta 811:265
43. Katz DJ, Stuchebrukhov AA (1997) J Chem Phys 106:5658
44. Daizadeh I, Gehlen JN, Stuchebrukhov AA (1998) J Chem Phys 109:4960
45. Ivashin N, Källebring B, Larsson S, Hansson Ö (1998) J Phys Chem B 102:5017
46. Voityuk AA, Rösch N, Bixon M, Jortner J (2000) J Phys Chem B 104:9740
47. Rust M, Lappe J, Cave RJ (2002) J Phys Chem A 106:3920
48. Matyushov DV, Voth GA (2000) J Phys Chem A 104:6470
49. Creutz C, Newton MD, Sutin N (1994) J Photochem Photobiol A Chem 82:47
50. Voityuk AA, Rösch N (2002) J Phys Chem B 106:3013
51. Voityuk AA, Jortner J, Bixon M, Rösch N (2000) Chem Phys Lett 324:430
52. Larsson S (1981) J Am Chem Soc 103:4034
53. Löwdin PO (1963) J Mol Spectrosc 10:12
54. Ratner MA (1990) J Phys Chem 94:4877
55. Skourtis SS, Beratan DN (1999) Adv Chem Phys 106:377
56. Priyadarshy S, Risser SM, Beratan DN (1996) J Phys Chem 100:17678
57. Steenken S, Jovanovich SV (1997) J Am Chem Soc 119:617
58. Rodrigues-Monge L, Larsson S (1996) J Phys Chem 100:6298
59. Hunter CA, Lu XJ (1997) J Mol Biol 265:603
60. Lu XJ, El Hassan MA, Hunter CA (1997) J Mol Biol 273:681
61. Clowney L, Jain SC, Srinivasan AR, Westbrook J, Olson WK, Berman HW (1996) J Am Chem Soc 118:509
62. Bloomfield VA, Crothers DM, Tinoco I (1999) Nucleic acids: structures, properties, and functions, University Science Books, Sausalito
63. Kelley SO, Barton JK (1998) Chem Biol 5:413
64. Baik MH, Silverman JS, Yang IV, Ropp PA, Szalai VA, Yang W, Thorp HH (2001) J Phys Chem B 105:6437
65. To illustrate the weak effect of electron correlation on the electronic coupling for hole transfer, we mention HF and CASPT2 results for an ethylene dimer at an intermolecular distance of 3.5 Å, 0.571 eV, and 0.555 eV, respectively; cf. Ref. 58

66. Ridley JE, Zerner MC (1973) Theor Chim Acta 32:134
67. Dewar MJS, Thiel W (1977) J Am Chem Soc 99:4899
68. Dewar MJS, Zoebish EG, Healy EF, Stewart JJP (1985) J Am Chem Soc 107:3902
69. Stewart JJP (1989) J Comput Chem 10:209
70. Thiel W, Voityuk AA (1996) J Phys Chem 100:616
71. Voityuk AA, Zerner MC, Rösch N (1999) J Phys Chem A 103:4553
72. Voityuk AA, Rösch N (unpublished results)
73. Jortner J, Bixon M, Voityuk AA, Rösch N (2002) J Phys Chem A 106:7599
74. Rak J, Voityuk AA, Marquez A, Rösch N (2002) J Phys Chem A 106:7919
75. Dapprich S, Frenking G (1995) J Phys Chem 99:9352
76. Bagus PS, Hermann K, Bauschlicher CW (1984) J Chem Phys 80:4378
77. Nakatani K, Dohno C, Saito I (2000) J Am Chem Soc 122:5893
78. Lewis FD, Calgutcar R, Wu Y, Liu X, Hayes RT, Miller SE, Wasielewski MR (2000) J Am Chem Soc 122:2889
79. Lewis FD, Wu Y, Liu X, Hayes RT, Letsinger RL, Greenfield SR, Miller SE, Wasielewski MR (2000) J Am Chem Soc 122:12346
80. Tavernier HL, Fayer MD (2000) J Phys Chem B 104:11541
81. Siriwong K, Voityuk AA, Newton MD, Rösch N (2003) J Phys Chem B 107:2595
82. Davis WB, Hess S, Naydenova I, Haselsberger R, Ogrodnik A, Newton MD, Michel-Beyerle ME (2002) J Am Chem Soc 124:2422
83. Steenken S (1997) Biol Chem 378:1293
84. Carra C, Iordanova N, Hammes-Schiffer S (2002) 106:8415
85. Ladik J, Ye YJ (2002) Int J Quantum Chem 90:838
86. Hjort S, Stafström S (2001) Phys Rev Lett 87:228101
87. Endres RG, Cox DL, Singh RRP, Pati SK (2002) Phys Rev Lett 88:16601
88. Liao JL, Voth GA (2002) J Chem Phys 116:9174
89. Shafirovich VY, Dourandin A, Luneva NP, Geacintov NE (1997) J Phys Chem B 101:5863
90. Schwögler A, Burgdorf LT, Carell T (2000) Angew Chem Int Ed 39:3918
91. Lewis FD, Liu X, Miller SE, Hayes RT, Wasielewski MR (2002) J Am Chem Soc 124:11280
92. Voityuk AA, Michel-Beyerle ME, Rösch N (2001) Chem Phys Lett 342:231
93. Cai Z, Li X, Sevilla MD (2002) J Phys Chem B 106:2755
94. Al-Jihan I, Smets J, Adamowicz L (2000) J Phys Chem A 104:2994
95. Russo N, Toscano M, Grand A (2000) J Comput Chem 14:1243

Top Curr Chem (2004) 237:73–101
DOI 10.1007/b94473

Polarons and Transport in DNA

Esther Conwell

Department of Chemistry, University of Rochester, Rochester, NY 14627, USA
E-mail: conwell@chem.rochester.edu

Abstract It has been widely considered that the wavefunction of an extra electron or hole on the base stack of a DNA molecule is confined to a single site, i.e., base or base pair. There is, however, a theorem that an extra carrier on a one-dimensional chain minimizes its energy by forming a large polaron, its wavefunction extended over a number of sites. Thus it is expected, and calculations show, that the wavefunction of an extra electron or hole on DNA is delocalized, even for an arbitrary base sequence. The hole wavefunction is centered on a guanine (G) because the HOMO of G is higher by many kT than that of the other bases. To understand the significance of this and how it affects transport, we begin by reviewing the experimental data on electron and hole diffusion in solution and drift in an electric field in air or vacuum. The finding of Giese's group, that a hole moves by tunneling between Gs when there is a random sequence of bases with Gs within four sites of each other, can be reinterpreted as tunneling of the hole polaron from a position where it is centered on one G to a position where it is centered on another G. Further experiments show that if a G is followed by a series of four or more adenine–thymine pairs (A:Ts), a hole has some probability of hopping from G onto the bridge of As with thermal energy and then moving rapidly along the As almost unattenuated. Consistent with the idea of the hole being localized to one site, it was suggested that the motion along the As is nearest-neighbor hopping. We suggest that the motion along the As is polaron drift. Some evidence for this is the experiment of Kendrick and Giese which shows that a hole introduced on an A migrates through an $(A{:}T)_n$ sequence in a manner independent of n. We present a number of additional arguments, some based on excitons and exciplexes in DNA, that an extra hole or electron propagates as a large polaron. The properties and the motion of a large polaron in DNA are calculated with a simple tight-binding model. Additional evidence that an extra hole is not localized to one site comes from the excellent agreement with experiment of our calculations based on the polaron model of the relative trap depths of G, GG, and GGG traps. The large polaron model can also account for the switch from tunneling to on-bridge propagation as the number of A:Ts between the Gs goes beyond three. Including in the calculations the solvent and counterions makes little change in the extent of the polaron but greatly increases its binding energy and has a strong drag effect on its mobility. For DNA in air or vacuum propagation may also be by large polarons, but more must be known, e.g., about disorder in a low-humidity environment and how the negative charge on the backbone is compensated, before a definite conclusion can be reached.

Keywords Large polarons · Diffusion · Conduction · Hole traps · Solvation

1 Introduction

The DNA molecule has a polymer backbone consisting of two helical chains with alternating sugars and phosphates. Attached to the two helices are the bases: guanine (G), cytosine (C), adenine (A), and thymine (T). Hydrogen bonds pair G on one helix with C on the other, while A is paired with T. The bases are planar heterocycles. In the form of DNA found in vivo, i.e., in solution with water and appropriate concentrations of positive ions to balance the negative charge on the backbone, the bases are parallel to each other and spaced 3.4 Å apart, if thermal motion is neglected. This is the van der Waals distance, indicating strong interaction between π electrons on adjacent bases.

Not long after the structure of DNA was understood, it was suggested that conduction was possible along the base stack [1]. Because optical absorption indicates a HOMO–LUMO gap $\geq$4 eV, it is expected that no free carriers are created by thermal excitation unless the DNA is doped. Little success has been attained so far in doping DNA; this will be discussed in Sect. 3. Free carriers have been created by high-energy photons, leading to photoconductivity [2]. This has not been much studied, however. Irradiation with x-rays has been shown to create free electrons and holes on the base stack. These are quickly trapped, but electron paramagnetic resonance studies show that they have mean free paths between 3 and 11 base pairs below 77 K [3, 4] and 30 base pairs above 150 K [4]. Irradiation does not lend itself, however, to obtaining detailed information on transport of free carriers. There is evidence, from studies of low-frequency absorption, for a background of highly localized, i.e., trapped, carriers in λ-DNA [5]. These may arise from disorder or impurities. They absorb microwaves or mm waves in hopping between traps. Studies of free carrier transport began in earnest after it was suggested that the DNA molecule is a highly conducting wire [6].

Transport in DNA is of biological interest because oxidation of a base and subsequent transport of the resulting hole may result in mutation and carcinogenesis [7]. It is also of interest for nanotechnology where, if DNA were a good conductor, it could be used for wires and its recognition properties put to use for assembly of components in nanocircuits [8].

1.1 Experimental Background

1.1.1 *Hole Motion in DNA in Solution*

Free carriers have been created by injection. In a frequently used scheme, an impurity, usually an acceptor, is intercalated or otherwise attached to the DNA molecule. A photon impinging on the acceptor creates an empty level into which an electron from one of the bases goes. This leaves behind a hole on the base stack. Diffusion of the hole on the stack has been studied intensively by a number of research groups. The progress of the hole is monitored mainly by one of two techniques: (1) quenching of the fluorescence of an intercalated fluorescent molecule; or (2) setting a series of partially transmitting traps, each trap usually consisting of a pair of Gs, GG, or three Gs, GGG, together. The number of holes caught at each trap could be determined by subsequent analytical work that detects the damage due to the trapped hole.

With these techniques many groups investigated the progress of holes injected into various DNA sequences. The results were usually stated in terms of the parameter β defined by assuming the measured rate constant k varies with the distance R traveled by the hole according to $k \propto \exp(-\beta R)$. The distance dependence predicted for electron or hole transfer by superexchange tunneling is given by the latter expression [9]. The values of β reported by various investigators ranged from 0.1 to 1.4 Å^{-1}. Finally the values converged to the range between 0.6 and 0.8 Å^{-1} for situations where the hole encountered only a few A:Ts between its injection and trapping sites. These values, although smaller than the values for transport in proteins, appear reasonable for single-step superexchange tunneling in DNA [9].

An explanation for the very low values of β found by some groups came from experiments in which the sequence was varied. Although the charge transfer nearly vanished when the injection site and the trap were separated by four A:Ts, replacement of the second or third A:T by a G:C base pair was found to increase the rate of hole transfer by 2 orders of magnitude [10]. It was concluded that the actual transport process in a DNA with mixed bases is tunneling from one G to the next [10, 11], provided the next G is within three or four sites.

Although Giese's group found that few holes would get through four A:Ts between a pair of Gs, suggesting that essentially none would go through more than four A:Ts, this appeared to conflict with earlier data of the Barton [6] and Schuster [12] groups which showed that some holes could penetrate through many more than four A:Ts. Motivated to do further experiments, Giese's group found that a few holes could move rapidly through many A:Ts beyond the third essentially without additional attenuation [13]. They suggested that the time to tunnel through four A:Ts becomes comparable to the time required for the hole to acquire sufficient thermal energy to hop onto the bridge of As. Once on the bridge the hole can move freely, there being no chemical reaction with water of the type experienced by a guanine cation. It should be noted that the energy difference expected between the HOMO levels of A and G was ~0.5 eV, based on one-electron redox potentials of the nucleobases in solution [14, 15] and on values of their ionization potentials in vapors [16, 17]. From the fact that the holes could acquire the energy thermally to hop from G to the bridge of As, it is apparent that the energy difference between the HOMO levels is no larger than ~0.2 eV. This is only one manifestation of the fact that the ionization potential of a base can be much affected by its neighbors.

The experiments just discussed made it clear that the motion of the hole on the series of As represents a different mechanism of transport than tunneling. Giese [13] and Bixon and Jortner [18] suggested that this mechanism is incoherent hopping of the hole between neighboring bases. This means that the hole wavefunction is limited to one base. The wavefunctions of the remaining electrons on that base would of course be distorted by the presence of the hole. Thus in this view of the transport process the base on which the hole sits could be called a molecular polaron, or a small polaron because it is limited to one site.

It had been suggested earlier by Schuster [12] that the hole wavefunction and the accompanying distortion would be spread over a number of bases, making it a large polaron. However, Schuster envisioned the polaron moving by a process, termed phonon-assisted polaron hopping, in which the charge could hop from one set of bases to another set of different bases provided the ionization energies of the two sets were similar. This mode of motion could not explain the experimental finding of Giese's group of "hopping between Gs". As will be elaborated in the remainder of this chapter, we believe it is correct that the hole wavefunction is spread over a number of bases, making it a large polaron. However, we describe the unattenuated motion beyond three A:Ts as polaron drift, to be described in Sect. 2, interrupted presumably by occasional scattering processes. It may be noted that, on the basis of a more recent experiment, to be discussed in the next section, Kendrick and Giese have suggested that a hole is not localized at one site but may be spread over several bases [19].

1.1.2 *Electron Motion in DNA in Solution*

Only quite recently have data become available on electron motion in DNA in solution. It was pointed out earlier that transport of negative charge would be likely to involve both T and C because they are the most easily reduced bases and they have similar reduction potentials [20]. With one of these present in each base pair, it can be expected that electron transfer should be relatively facile compared to hole transfer and, in addition, much less dependent on sequence.

These ideas were tested in an experiment in which a photoexcited reduced flavin acted as a donor, and a thymine dimer, separated from the flavin by a number of A:Ts, as an acceptor. The thymine dimer was modified so that arrival of the electron caused a strand break, allowing the dimer to serve also as a detector [21]. Varying the number of A:Ts between the flavin donor and dimer acceptor, Carell's group found that, for this system, electrons traversed the A:Ts with much less attenuation than had been found for holes in the previously investigated systems [11]. Each A:T decreased the number of electrons by only about 30% [21], whereas for the holes the decrease per A:T, up to four, was a factor of ~10. The data led to a β value of 0.1/Å, much too small for tunneling. Carell et al. suggest that the transport mechanism is nearest-neighbor hopping, just as in the case of holes in a long sequence of As. We suggest, however, that in this case also the motion could be described as polaron drift.'

1.1.3
Conduction in DNA in Air or Vacuum

Current in response to a dc electric field has been observed in a number of experiments with the DNA in air or vacuum. The electric field measurements produced a wide variety of conductivities, DNA being characterized at one extreme as highly insulating [22, 23], at the other extreme as highly conducting or metallic, becoming a superconductor at low temperatures [24]. It should be noted that the experiments in which high conductivity was found [24, 25] were carried out on long (up to ~1 μm) λ-DNA molecules. These molecules have a random DNA sequence, which means that, due to the relatively large differences in HOMO levels of the bases, holes would inevitably encounter large barriers to their motion. Electrons would encounter smaller barriers but it is quite unlikely, on the basis of the data of the Carell group, that electron transport could lead to such high conductivity in a long molecule with random base sequence, such as λ-DNA. It can only be concluded that the low resistivities found were the result of inadvertent doping or a conducting surface layer. The low resistivity found by Fink and Schonenberger [25] can be attributed to the latter; de Pablo et al. [22] showed that slow electron bombardment of the kind to which the samples were subjected in the experiments of [25] built up a low-resistance contamination layer on the surface. Rakitin et al. claimed to observe conductivity in λ-DNA, although much lower than reported in [24] and [25], but their DNA consisted of an estimated 3×10^2 strands [26].

To avoid the barriers due to different HOMO and LUMO levels on different bases, many experimenters used DNAs made up of all the same base pairs, G:Cs or A:Ts. Carriers were introduced by injection at the contacts. Not unexpectedly, it was found that poly(dG)-poly(dC) exhibits p-type behavior, i.e., hole transport, while poly(dA)-poly(dT) exhibits n-type, i.e., electron transport [27]. The experiments on DNA with all the same base pairs produced, for the most part, moderate resistivity values, characteristic of a semiconductor. Here again, however, different experimenters, working on a single duplex strand of DNA in air or vacuum with all pairs the same, obtained quite different results. Porath et al., working on 10.4-nm oligomers with repeated G:C, found a voltage threshold for measurable current of 2 to 3 V at low temperatures, increasing to 4 V or more at 300 K [28]. They concluded that most of the gap probably arises from the offset between the Fermi level of the electrodes, platinum in their case, and the molecular energy bands of the DNA. Yoo et al. found a similar voltage threshold at low temperatures but, in contrast to the results of Porath et al., their threshold decreased with increasing temperature, vanishing at room temperature [27]. Even at low temperatures the current found by Yoo et al. for poly(dA)-poly(dT) at a given voltage beyond the threshold is comparable to that found by Porath et al. for poly(dG)-poly(dC), which is surprising because the Yoo et al. samples are ~0.5 to 1.5 μm in length, about 100 times as long as the Porath samples. Even more surprising, at room temperature they found the resistivity of poly(dG)-poly(dC) to be 0.025 Ω cm. This is also quite in contradiction to the results of [23], where long poly(dG)-poly(dC) samples were found to have extremely high resistivity at room temperature. Of course, we are dealing with injected rather than intrinsic carriers in these experiments, but to achieve such low resistivity with injected carriers is quite unlikely, except perhaps in cases where an additional electrode is used, as in a field-effect transistor.

Yoo et al. found that for poly(dA)-poly(dT) their measured current varies linearly with sinh (bV), where b is a constant independent of temperature [27]. They point out that for a small polaron hopping model [29] the current is predicted to depend on the voltage V according to sinh $(eaV/2kTd)$, where e is the charge on the electron, a the hopping distance, k the Boltzmann constant, T the absolute temperature, and d the distance between the electrodes. To make the behavior of poly(dA)-poly(dT) fit this small polaron hopping model it must be assumed that a increases linearly with T, rather than being constant as assumed in deriving the model. For poly(dG)-poly(dC), however, a fit could be obtained with the hopping distance a independent of T. No explanation was offered for the difference between the two different sequences.

Transport in DNA samples with all bases the same could be either by free carriers, i.e., band transport, or by polarons. As will be further discussed in the next section, the polarons are expected to be large polarons, not small. In the conducting polymers there is overwhelming evidence that electrons (holes) from a metal contact are injected directly into polaron states in the polymer, because the polaron states have lower energies than the LUMO (HOMO) or conduction (valence) band edge. As has recently been shown theoretically [30], the injection takes place preferably into a polaron state made available when a polaron-like fluctuation occurs on the polymer chain close to the interface, rather than into a LUMO state, with subsequent deformation to form the polaron. It could also be expected for DNA that injection takes place into a polaron-like fluctuation whether the injection is directly from a contact or from a G onto a bridge of As.

1.2 Why Large Polarons?

In this section we present the arguments for the wavefunction of an excess electron or hole on the DNA base stack to be extended over a number of bases. This will serve also as an introduction to the detailed discussion of polaron properties to be taken up in Sect. 2.

It has been shown theoretically that an extra electron or hole added to a one-dimensional (1D) system will always self-trap to become a large polaron [31]. In a simple 1D system the spatial extent of the polaron depends only on the intersite transfer integral and the electron–lattice coupling. In a 3D system an excess charge carrier either self-traps to form a severely localized small polaron or is not localized at all [31]. In the literature, as in the previous sections, it is frequently assumed for convenience that the wavefunction of an excess carrier in DNA is confined to one side of the duplex. This is, of course, not the case, although it is likely, for example, that the wavefunction of a hole is much larger on G than on the complementary C. In any case, an isolated DNA molecule is truly 1D and theory predicts that an excess electron or hole should be in a polaron state.

A strong argument for transport by polarons is the very small attenuation found when a hole travels through a long series of As [6, 12, 13], or an electron travels between a flavin donor and a dimer acceptor [21]. Comparably small attenuation is found for a bridge consisting of a conjugated polymer when there is a small energy difference between the donor level and the bridge [32]. It has been very well documented, by many different kinds of experiments, that an excess charge on a conjugated polymer forms a large polaron and propagates as such with very little resistance or attenuation so long as it drifts on the same conjugation length [33].

An argument that has been given against polaron formation in DNA is that the thermal motion is so strong at room temperature that it would act as the rate-limiting factor for charge transfer between neighboring bases [34]. This would essentially nullify the effect of the finite transfer integral, limiting the wavefunction of an excess carrier to one site. A theory by Bruinsma et al. [34] based on this argument predicts that carrier mobility would increase strongly with increasing temperature. The Porath et al. data, taken for a temperature range from 4 K to room temperature, for voltages beyond the gap do not show such an increase [28].

It is noteworthy also that excitons, which have also been assumed, on the basis of some of the optical properties of DNA, to be localized to one base, have recently been shown to be delocalized. In the experiments of Rist et al. [35] it has been shown that, when there are two identical bases adjacent to each other, the exciton is delocalized over the two bases.

The theory of Bruinsma et al. [34] was motivated by experimental results of a group including Barton and Zewail [36]. In the experiment an intercalated molecule of an acceptor, ethidium (E), and that of a donor molecule, deazaguanine (Z), were incorporated into a DNA molecule. The position of E was fixed but that of Z varied so that there were three, four, or five bases between acceptor and donor. E was excited into its singlet state and the dynamics of the excited state decay studied by means of transient absorption and fluorescence upconversion. Time constants of 5 and 75 ps were observed to characterize the decay, independent of the distance between E and Z [36]. It was assumed that these two time constants represented the time for hole transfer between E and Z. The fact that the time constants were so long was attributed to the dynamical nature of the DNA base stack giving rise to a distribution of conformations at any given time, of which only a fraction were suitable for electron transfer [36]. As emphasized by Lewis, however [37], it was never verified in the experiments that a hole actually traveled from E to Z. In studies of the singlet-state decay of several hairpins where there was no donor to travel to, Lewis showed that the decay included time constants similar to those found in [36]. He attributed these time constants to excited-state relaxation processes of the singlet [37]. This explanation for the time constants observed in the decay of the E singlet would also account for the lack of distance dependence observed for the decay in [36].

Although the considerations of the last paragraph remove what could have been a rather direct proof that the thermal motions limit the wavefunction of an excess electron or hole to one site and thus determine transport in DNA, it is still true that various experiments indicate strong thermal motion on a ps time scale [38], and molecular dynamics calculations show large changes in the structure of DNA on a ps-ns scale [39]. An argument against this motion causing localization of an excess hole wavefunction to one site is that thermal motions in the conducting polymers, such as polyparaphenylene, should be still more vigorous at room temperature because the atoms that are the sites are very much lighter. Yet, as noted earlier, there is overwhelming evidence that excess charges in the conducting polymers form polarons and propagate as polarons. A significant reason for this is that the polaron structure acquires more rigidity or stability to dampen relative motions of the molecular constituents that would tend to increase the polaron energy by decreasing the π overlap. In the case where the molecular constituents are rings, e.g., polyparaphenylene, in the presence of the excess charge the five rings over which the charge is spread tend to become quinoid [40], thus increasing the rigidity of the polaron. Because the geometry is different, this particular bonding change would not occur in DNA, but it is reasonable to expect changes in bonding that would tend to reduce those relative motions of the molecular constituents that decrease the π overlap.

It is instructive to consider the stability of other excitations in DNA, excimers and exciplexes. An excimer (exciplex) is formed when two identical (nonidentical) molecules that do not interact in their ground states do so when one of the molecules is in an excited state. As a result of charge-transfer and exchange interactions of the overlapping π electrons of the two molecules on the one hand, and their mutual repulsion on the other, the molecules are drawn together in a potential minimum at a separation smaller than that in the ground state [41]. These features result in the emission from the excimer or exciplex being red-shifted, broad and featureless in comparison to that from a singlet exciton. Singlet excimers and exciplexes are more common than triplet, and the former usually consist of the two molecules arranged in a sandwich fashion [41]. The separation of the two molecules in the excimer or exciplex is typically 3 to 3.5 Å. Note that from the above description the likely geometry of the singlet excimer or exciplex and its formation are quite similar to those of the polaron.

A number of excimers, notably adjacent C-C and A-A, and exciplexes consisting of adjacent A-C and A-T, have been identified at 78 K [42]. Only A-T emits at room temperature [43], however, the emission being seen from poly(A:T)-poly(A:T). Significantly, the lifetime of this fluorescence at room temperature is ~7 ns [44]. Of course, this is the radiative lifetime, thus a lower limit on the possible lifetime of the excitation. Thus, despite the vigorous thermal motion, this excitation, which involves a distortion quite similar to that of a polaron, lasts at least nanoseconds.

In the experiment of Kendrick and Giese [19] referred to earlier, a hole is injected into an A. On one side of the A is a sequence of two A:Ts followed by a GGG trap, while on the other side is a sequence of eight A:Ts followed by a GGG trap. As judged by the subsequent yields of strand-cleavage products at the traps, the efficiency of the hole migration through the A:T sequences to the trap depends very little on the number of As between the location of the injection and the trap. This is an unlikely outcome if the charge transport were by incoherent hopping. Kendrick and Giese suggest that a possible explanation is that the hole injected into the adenine is delocalized over adjacent A:T base pairs [19].

In the next section we will discuss in greater detail the structure and evidence for large polarons. In Sect. 2.1 we present our model for calculating the properties of the large polaron and a discussion of the parameters used in the model. It will be shown in Sect. 2.2 that this model leads to very good agreement with experimental results on relative depths of G, GG, and GGG traps, in contrast to the model based on the hole being localized on a single base. In Sect. 2.3 a stability analysis is carried out to determine how an excess hole will move, i.e., whether it will tunnel or propagate on the bridge, on a stack with a series (bridge) of As separating a single G and a GGG trap, as in the experiments of [11]. Comparison of our results from the polaron model with the results of [11] gives further evidence for the polaron mechanism. In Sect. 2.4 we consider the effects of the environment, notably the water and ions surrounding the DNA. Polarizability of the surrounding medium, although making little change in the length of the polaron, greatly increases its binding energy.

Turning to the calculations of polaron mobility in Sect. 2.5, we find that, although a stationary polaron can form with the wavefunction extending over an arbitrary sequence of bases, in the absence of an electric field, or in a small electric field, the polaron cannot move far unless the DNA is made up of the same base pair repeated. This result is for zero temperature, of course, not allowing thermal energy that makes possible the transition discussed earlier from G to A. With that exception the result is in agreement with experiment, as discussed earlier. When all the bases are the same the polaron can move by drifting, with shape unchanged, due to small motions of successive bases. With a large field our calculations show that the polaron can surmount barriers of tenths of an eV. Using the results of Sect. 2.4, we calculate the drag effect of the water on polaron motion, obtaining upper limits on the diffusion rate and mobility for holes in a DNA molecule in solution. For nanotechnology applications it would be desirable to have higher conductivity DNA. Sect. 3 is devoted to a discussion of what has been achieved so far by doping. In Sect. 4 we comment briefly on some of the recent calculations of DNA conductivity in the literature.

2 Properties of Polarons in DNA

2.1 Calculations for a Stationary Polaron

The essential ingredients of a Hamiltonian that gives rise to polarons are: (1) a term for hopping or transfer between neighboring sites; (2) dependence of the hopping term on the spacing of the neighbors, which may change due to motions of the sites; and (3) an elastic restoring force that keeps the distortion due to motion of the sites finite.

In our calculations of polaron properties in DNA we have assumed only one degree of freedom, longitudinal motions that change the spacing of the bases. If the change in spacing is small enough, the hopping or transfer integral t may be expressed as $t=t_0-\alpha(u_{n+1}-u_n)$ where t_0 is the transfer integral when every pair of sites is separated by the same distance a, u_n is the displacement of the nth site, and $\alpha\equiv\partial t/\partial u$. A Hamiltonian that incorporates this feature and satisfies the criteria given above is the Su–Schrieffer–Heeger (SSH) Hamiltonian [45]. It has been used successfully to describe the properties of polarons in conducting polymers. Written for a situation in which there may be sites of different energies, and allowing for the addition of an electric field, this Hamiltonian becomes:

$$H = H_{\mathrm{el}} + H_{\mathrm{lat}} \tag{1}$$

$$H_{\mathrm{el}} = \sum_{n,s} d_n c^+_{n,s} c_{n,s} - (t_0 - \alpha(u_{n+1} - u_n)) \left[e^{i\gamma A} c^+_{n,s} c_{n+1,s} + e^{-i\gamma A} c^+_{n+1,s} c_{n,s} \right] \tag{2}$$

$$H_{\mathrm{lat}} = {}^{K}\!/_{2} \sum_n (u_{n+1} - u_n)^2 + {}^{M}\!/_{2} \sum_n \dot{u}_n^2 . \tag{3}$$

Here C^+_n is an electron or hole creation (annihilation) operator at site n, d_n are on-site energies, including the polarization energy, K the spring constant, M the mass at each site, $\gamma\equiv 2\pi ea/hc$, c being the light velocity, e the electronic charge, and A the vector potential of the external electric field.

The spin indices have been dropped because the quantities are not spin-dependent. Equation 3, embodying a classical treatment of the lattice vibrations, is justified by the large mass of the sites, the bases in the case of DNA. We have used this Hamiltonian, and variations on it, to calculate the properties of polarons in DNA.

Static solutions are found by minimizing the Hamiltonian above with respect to the nearest-neighbor distance change $y_n = u_{n+1} - u_n$ and the electronic energies, which yields the following equations for the hole polaron:

$$(E_h - 2t_0)\psi_n = d_n\psi_n - (t_0 - \alpha y_n)\psi_{n+1} - (t_0 - \alpha y_{n-1})\psi_{n-1} \tag{4}$$

$$y_n = -(2\alpha/K)\psi_{n+1}\psi_n \tag{5}$$

where $E_h - 2t_0$ is the energy of a hole relative to the bottom of a free hole band in a chain with d_n and u_n both equal to zero. Note that only the strain $u_{n+1} - u_n$ enters the equations, not the displacements themselves. The total energy of the system in the stationary state is taken to be the sum of the hole energy obtained from Eq. 4 and the lattice deformation energy, the first term in Eq. 3.

These equations must be solved numerically. In our first calculations [46] the value of t_0 was taken as 0.3 eV, the value obtained from an ab initio calculation of the energy of a pair of Gs as a function of the distance between them [47] for a separation of 3.4 Å. The electron–phonon coupling constant α, the derivative of t with respect to displacement, was obtained as 0.6 eV/Å from the results of [47]. Subsequently, when a value of t_0 other than 0.3 eV was used α was scaled accordingly. Although t_0 and α values depend on the particular pair of neighboring bases, the calculations were simplified by using the same value for all pairs of bases. The value of the elastic constant K was taken as 0.85 eV/Å^2, derived [46] from the measured value of the sound velocity in DNA.

In Fig. 1 we show the properties of the polaron calculated for the sequence shown at the bottom of the figure. The lattice strain (always negative) is represented in the figure by solid symbols placed at half-integer abscissas ν=..., −1/2, 1/2, 3/2,..., which correspond to the differences $u_{\nu+1/2} - u_{\nu-1/2}$. To obtain the results shown in Fig. 1 we took d_n=0 for G, and 0.17 eV for A in Eq. 2. As noted in Sect. 1.1.1, because a hole can jump from G to A by means of thermal energy, the difference between the site energies of A and G is no more than 0.2 eV. The choice of this difference as 0.17 eV will be discussed in the next section. Solving for the wavefunction and energy of the polaron for many different sequences, using the values of ionization potentials to obtain on-site energies for C and T, we found that the extent of the polaron, about six sites in Fig. 1, was much the same for different sequences [46]. It is seen in Fig. 1 that change in t_0 changes only the strain (change in nearest-neighbor distance). The maximum strain is ~0.3–0.4 Å. The calculated binding energy was 0.187 eV for t_0=0.3 eV, 0.110 eV for t_0=0.2 eV [48]. It will be seen in Sect. 2.4 that the binding energy is greatly increased by the DNA being in solution.

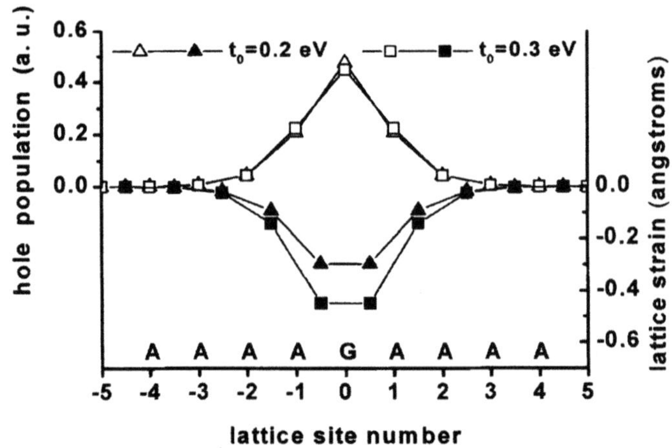

Fig. 1 Hole population, $|\psi_n|^2$, (*open symbols*) and lattice strain (*filled symbols*) for one guanine at n=0 among adenines: t_0=0.2 eV (*triangles*) and t_0=0.3 eV (*squares*) (From Conwell and Basko [48])

Since our earliest calculations were done, there have been a number of calculations of t_0 [49–52]. Although the results have varied widely, the t_0 values have in general been smaller than 0.3 eV, frequently ~0.1 eV or less. To explore the effect of variations in the structure, Voityuk et al. calculated the coupling matrix elements for hole transfer between adjacent Watson–Crick pairs in DNA for different values of the three possible translations and the three possible rotations of the pairs [50]. They found that the coupling is very sensitive to variations of the relative positions of the pairs. On this basis the effective t_0 should reflect not only the effect of relative translations of the bases but also the other motions that affect the magnitude of the coupling. This speaks for a larger value of t_0 than obtained taking only translational motion into account. Taking into account other motions could also affect the geometry of the polarons, but that is a secondary effect.

There is another argument for a larger value of t_0 to be used in our calculations. Voityuk et al. found that the electronic coupling between a pair of A:Ts varied, in the range 2.38 to 4.38 Å, as $\exp[-2.0(R-3.38)]$, where R is the actual separation in Å [50]. Grozema et al. found that the coupling between a pair of Gs varies as $\exp(-aR)$, where a=1.7 Å^{-1} [52]. For t_0=0.1 eV, we can estimate from Fig. 1 that the maximum strain, or change in spacing, due to creation of the polaron is ~0.2 Å. If the change in t varied linearly with spacing, as assumed in Eq. 2, with α scaled for the smaller t_0 value a decrease in spacing of 0.2 Å would increase t by 0.024 eV. With the exponential rate of increase calculated by Voityuk et al. the change in t due to a decrease in spacing of 0.2 Å would be an increase by 0.15 eV. It appears that, on average, the polaron sees a larger value of t than given by the linear rate of increase assumed in the calculations based on the SSH Hamiltonian. Thus it is not unreasonable that the situation is better represented by a higher value of t_0 than is found in calculations of the coupling matrix element for the equilibrium spacing of the bases, 3.4 Å. With the numbers just given, and the argument that motions other than translation contribute to the coupling, a t_0 value of 0.2 eV seems a good choice.

2.2
GG and GGG Traps

As mentioned earlier, the motion of a hole on DNA has been studied by observing the fraction of holes trapped at a series of partially transmitting traps, commonly GG or GGG. To detect the trapping, use is made of the fact that the radical cation may react irreversibly with water or oxygen resulting, with some further chemical treatment, in cleavage of the DNA at the site of the trapped hole. It is considered that the relative reactivity of a hole trap can be determined by densitometric assay of the cleavage bands seen in high-resolution polyacrylamide gel electrophoresis. Measurements of relative reactivity made with this technique show that the GGG traps are more reactive, although not greatly so, than GG. For a duplex DNA containing a G, a GG, and a GGG, Hickerson et al. report a cleavage ratio of 1:3.7:5.3 [53]. These numbers could vary somewhat, depending on the surrounding bases, but many other determinations found similar low cleavage ratios.

Two different explanations have been advanced for the difference in trapping rates of GG and GGG. Berlin, Burin, and Ratner [54] base their explanation on the assumptions that: (1) both traps are deep, i.e., ~0.5 eV for GG and ~0.7 eV for GGG, and (2) the ionization potential of guanine is lower than that of the other nucleobases by at least 0.4 eV. The different reactivities of the two traps were ascribed in [54] to different relaxation times of a hole in the trap. The GG units were taken to have a long relaxation time, so that a hole is likely to make a further hop before the trap closes on it, while the relaxation time of the GGG units was supposed to be relatively short, faster than the hopping time.

The assumptions about the trap depths were based on data and calculations for isolated bases in solution [14–17, 47]. As emphasized by Schuster [12], among others, there are many types of evidence that the ionization potential of a base can be much affected by its neighbors. Among the evidence is the fact that ab initio calculations of the ionization potentials of pairs or triples of stacked bases show considerable differences from the single base values [47, 55]. The trap depths were measured by Lewis et al. [56] with experiments on synthetic hairpins that included on a strand GG or GGG units among As. They found that the free energy liberated in a hole transfer from G^+ to a GG is 0.052 eV, while hole transfer from G^+ to a GGG resulted in a free energy of 0.077 eV. Thus the traps are fairly shallow. These numbers suggest also that the stronger reactivity of GGG can be accounted for by its greater depth. Our contribution was to show that excellent agreement with the experimental depths is obtained when the bound hole is treated as a polaron, as will be detailed below. Calculations in which the hole was assumed to be confined to a single site led to the difference in energy between G^+ and

$(GG)^+$ or $(GGG)^+$ of 0.3–0.13 eV, depending on the surrounding bases [57], in poor agreement with the experimental values.

For the calculations we used Eqs. 1–5. These equations neglect the effects of the environment which, we have indicated, are significant. Nevertheless, we believe they can be neglected here because we are calculating the difference of two energies, in which the system differs only in the replacement of one or two As by Gs. For simplicity the hole is assumed to be confined to a single strand of the DNA duplex. Because we are studying hole trapping on guanines placed in a chain of adenines, it is convenient to set the quantity d_n=0 on the adenine sites and equal to the difference between the ionization potential of G and that of A, to be denoted Δ_{GA}, on the guanine sites. We discovered that the results are very sensitive to the value of Δ_{GA}; it was therefore taken as a parameter in the calculations. The calculations were carried out using Eqs. 1–5 with t_0 values of 0.1, 0.2, and 0.3 eV and corresponding α values of 0.2, 0.4, and 0.6 eV/Å, respectively.

To compare with the results of Lewis et al. [56] we carried out the calculations for sequences similar to those used by them. Figure 1 shows the results obtained for a single G among As with Δ_{GA}=0.17 eV. It was found that this value of Δ_{GA} led to excellent agreement with the measured energy differences between the traps. Therefore we report only results obtained with this value. The results for GG surrounded by As are shown in Fig. 2. The calculations were actually done for long series of As surrounding the Gs, whereas in the experiments the chains were much shorter. We show in this figure that there is little difference between the series AGGA and a series in which GG is surrounded by many more As. In Fig. 3 we show the results for GGG compared with those for G, both surrounded by As. It is seen that the differences between the wavefunctions and the strains for G and GGG are quite small, consistent with the energy difference between G and GGG being quite small.

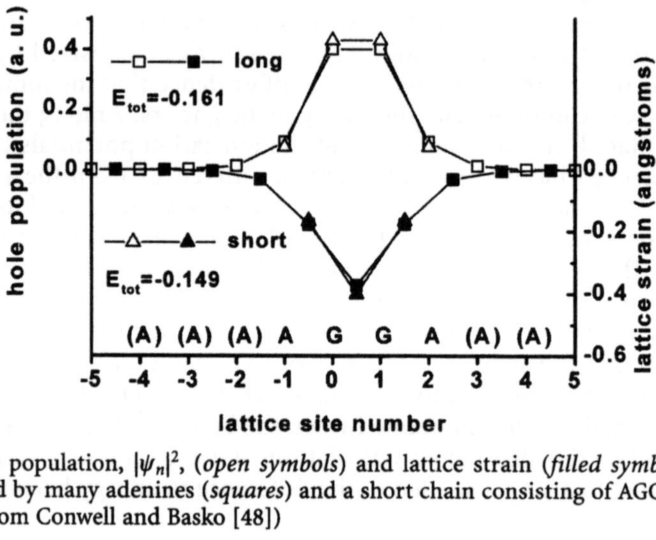

Fig. 2 Hole population, $|\psi_n|^2$, (*open symbols*) and lattice strain (*filled symbols*) for GG surrounded by many adenines (*squares*) and a short chain consisting of AGGA only (*triangles*) (From Conwell and Basko [48])

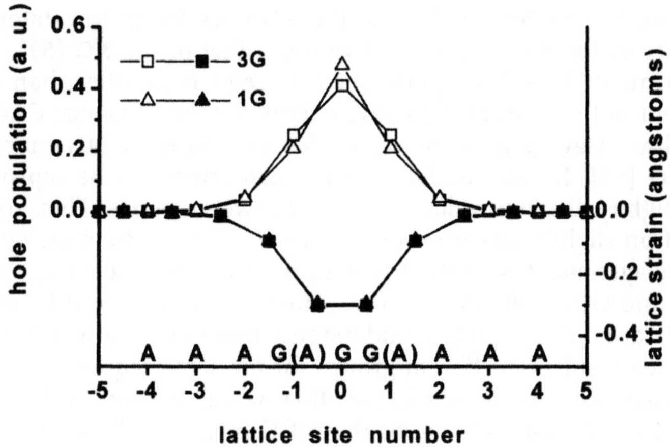

Fig. 3 Hole population, $|\psi_n|^2$, (*open symbols*) and lattice strain (*filled symbols*) for one guanine at n=0 (*triangles*) and three guanines at n=−1, 0, 1 (*squares*), in both cases surrounded by adenines (From Conwell and Basko [48])

The calculations gave the result that the energy released when a hole goes from G to GG is 0.051 eV for t_0=0.2 eV and 0.053 eV for t_0=0.3 eV. The experimental result of Lewis et al. was 0.052±0.006 eV. For a transition from G to GGG the calculated energy release was 0.078 eV for t_0=0.2 eV, and 0.081 eV for t_0=0.3 eV, while the experimental result was 0.077±0.005 eV [56]. For t_0=0.1 eV the values for the energy released were still within 10% of the experimental values. It is seen that the calculated values are in excellent agreement with the experimental values and insensitive to the value of the transfer integral. However, the strains are not insensitive to the value of t_0. For a t_0 value of 0.1 eV, which still fits the experimental values for the energy differences of the traps quite well, the maximum strain is 0.2 Å rather than 0.4 Å as obtained with t_0=0.3 eV.

The calculated values are, as noted earlier, quite sensitive to the difference in the ionization potentials of G and A. A change from 0.17 to 0.23 eV in Δ_{GA} results in a difference of 40% between the calculated and experimental results. As noted in Sect. 1.1.1, to be consistent with the observation that a hole can make the transition from G to A thermally, Δ_{GA} cannot be much larger than 0.17 eV.

Additional evidence for the correctness of our results is obtained from cleavage studies. From $|\psi_n|^2$ for GGG in Fig. 3 we obtain equal populations for sites n=1 and −1, with the population of the middle G, at site n=0, being 1.65 times as large. This matches surprisingly well the ratios seen in piperidine cleavage of the sequence AGGGA [58], which supports the usual assumption that the relative probability of cleavage on a site n in a GGG trap is proportional to the probability $|\psi_n|^2$ of finding the hole on it.

In our calculations the populations $|\psi_n|^2$ are necessarily equal for the two guanines in a GG trap, due to the symmetry assumed in the calculation. This would correspond to the cleavage ratio 1:1. Many determinations of the cleavage ratio for the 5′G to that for the 3′G have led to the statement that the reactivity of the 5′G is three to five times that of the 3′G [59]. However, these experiments have been performed on sequences other than AGGA. A measurement of the ratio of 5′ to 3′ reactivity for the sequence CAGGAT under piperidine cleavage gave the result that the 3′G is slightly more reactive than the 5′G [58]. In our modeling this small effect can be reproduced by ascribing slightly different values of d_n to the two guanines. This would make the population slightly larger on one of the Gs and can be done without destroying the agreement with the Lewis et al. results for the energy difference between G and GG. It should be noted, however, that the results of cleavage experiments are not completely understood, there being unexplained differences between results of different groups for the same sequence.

In summary, we have shown in [48] that we can account well for the trap depths of GG and GGG relative to that of G measured by Lewis et al. [56] with a model in which the wavefunction is not confined to the Gs but is still substantial on the surrounding bases, As in this case. The fit is insensitive to the value of the transfer integral, but requires that the difference between ionization potentials of adjacent G and A be ~0.2 eV rather than ~0.4 eV characteristic of the isolated bases. The small trapping found can be attributed entirely to the shallowness of the traps and, contrary to the assumption of [54], does not require different relaxation rates of the traps.

2.3 Transition from Tunneling to On-Bridge Propagation

In the experiments of Giese et al. mentioned earlier [11, 13], a hole is injected into a DNA chain in solution at a guanine site that is separated by a series of A:Ts from a GGG sequence. When the number of A:Ts is three or less, the fraction of holes reaching the GGG trap is reduced by a factor of ~10 for each intervening A:T. This indicates that the holes propagate by tunneling. Beyond three A:Ts, however, there is essentially no further attenuation of the holes as they propagate through many additional A:Ts, indicating a change in conduction mechanism. Giese et al. suggested that in the time it takes to tunnel beyond three As, a few holes can acquire sufficient thermal energy to hop onto the bridge of As [13]. Further progress of the holes was attributed by them to incoherent, nearest-neighbor hopping between As.

A theoretical study of the states of the system has been carried out to understand the motion of the hole from the shallower G trap or well, through the As, to the GGG deeper trap or well [60]. For each well there is a stationary state with the wavefunction of the hole localized around the well. Examples of these stationary states are shown in Sect. 2.2. The state in the deeper well represents the ground state of the system; the other is an excited state. In a geometric view of the situation, each possible wavefunction of the system, $\psi(x)$, may be thought of as a point in function space. Then the ground state of the system is the global minimum of the energy functional $E[\psi]$ in this space, while the excited state may be either a local minimum or an unstable stationary point. If the dynamics of the system were governed only by the equations of motion set out in Sect. 2.1, the system could remain in the excited state forever. However, if one includes fluctuations and dissipation, the character of the excited state becomes very important in determining further motion of the hole. For a local minimum small perturbations will not drive the system out of the stationary point (i.e., the hole stays in the shallow well), while in the case of an unstable point dissipation may make the system fall into the global minimum (i.e., the particle moves into the deep well).

Thus it is necessary to study the stability of the state in the G well. To obtain the energy functional it is convenient to start from the Hamiltonian given by Eqs. 1–3. Assuming a stationary solution $\psi_n(t)=\psi_n \exp(-iEt/\hbar)$, and substituting Eq. 5 into the Hamiltonian, we obtain the energy functional

$$E[\psi_n] = 2 + \sum_n \left[V_n \psi_n^2 - 2\psi_n \psi_{n+1} - (g/2)\left(\psi_n \psi_{n+1}\right)^2 \right] \tag{6}$$

where the energy and V_n are measured in units of t_0 and $g=4\alpha^2/Kt_0$. The constant 2 is added to set the zero of energy at the extremum of the valence band in an undistorted chain ($V_n=0$) because we are dealing with holes.

In the experiments of [11] the G and GGG traps were separated by a series of As that we take to be ℓ in number. For the calculation the traps were modeled by shifting down V_n for $n=0$, $\ell+1$, $\ell+2$, $\ell+3$ by an amount Δ, the difference between the ionization potentials of A and G in units of t_0. The difference between the ionization potentials was determined to be ~0.2 eV, as discussed in the last section, and t_0 was taken as 0.2 eV, which makes $\Delta=1$ for the case with which we are concerned. The task then is to find and compare saddle points of the energy functional. We look for saddle points of the functional (Eq. 6) numerically minimizing $\Sigma_n(\partial E/\partial\psi_n)^2$ and then determining the character of the saddle point by looking at the shape of the wavefunction corresponding to each point. For sufficiently long bridge length ℓ (such that the states in the two wells overlap weakly) there are two possible trajectories: (1) the hole jumps out of the G trap, travels across the bridge, and falls into the GG trap; and (2) the hole tunnels to the GG trap through the exponential tail of the wavefunction, a small fraction of the hole appearing first in GG and growing slowly until the wavefunction is entirely in GG, the wavefunction on the bridge being small throughout the process.

The results of these calculations for different values of Δ and ℓ are shown in [60]. For our case, with $\Delta=1$, the saddle points for tunneling and on-bridge propagation have different energies beyond $\ell=4$. For $\ell\leq 4$, however, the saddle points merge. According to the nature of the wavefunction, which has a peak in each well, the character of the saddle point in this range of ℓ is tunneling-like. Beyond $\ell=4$ for $\Delta=1$ the tunneling saddle point still has lower energy than the on-bridge propagation saddle point, but it must be realized that tunneling becomes unlikely if ℓ is large. Even for a very large separation, displacing an infinitesimal fraction of the polaron from G to GG may be energetically advantageous if the latter trap is deep enough. However, for large ℓ the characteristic time in which the instability develops, and most of the wavefunction leaks into the deeper trap, is determined by the tail of the wavefunction. Thus the time grows exponentially with ℓ in agreement with the Marcus–Levich–Jortner relation for the tunneling rate [61]. We conclude that, up to a critical bridge length, ℓ=4 by our calculations, the hole propagates by tunneling, whereas beyond ℓ=4 on-bridge propagation should dominate. Our result for the critical bridge length is in reasonable agreement with the experimental findings, where the switching between the two propagation mechanisms was observed at ℓ=3 [13].

The smaller value of ℓ found experimentally suggests that the coupling constant g is larger than we assumed. In our model of DNA we considered hole self-trapping due to interaction with longitudinal vibrations only. As discussed in Sect. 2.1, other degrees of freedom may also contribute to the coupling constant.

2.4 Environmental Effects—Solvation

So-called physiological conditions, i.e., those in which DNA is placed in nature, correspond to its being immersed in a 0.1 M solution of NaCl. Under these conditions DNA, being an acid, donates protons to the solution (one proton per phosphate group), which results in a negative charge on the backbone (−2e per lattice constant a=3.4 Å). A charge on the chain is subject to (1) dielectric screening by the surrounding water molecules, (2) Debye screening by the mobile counterions (Na^+ and Cl^-) with a characteristic distance (Debye screening length) ~10 Å, and (3) partial compensation by Na^+ ions that condense directly onto the DNA chain [62].

The problem we address in this section is the effect of this environment on the motion of an excess hole on a DNA molecule. There has been a recent calculation that focussed on the effect of the DNA environment, specifically on the effect of finite-temperature (configurational) dynamics of the hydrated counterions on the excess hole motion. This work concluded that there is a gating effect due to Na^+ ions close to the negatively charged phosphates or residing in the grooves of the DNA helix [63]. Our concern here is rather different; it is the effect on the hole motion of its polarization of the surrounding medium. The configuration of the solvent produced by the interaction with the static charge on the phosphate groups on average produces a constant shift of the energy of the hole, the same for all the bases. Obviously this will not affect the motion of the hole. What must be taken into account is the change in this configuration due to the presence of the hole, and the corresponding feedback effect on the hole motion. We assume that the presence of the hole does not affect the counterions condensed on the chain. Even if the hole were completely localized on one base it would bring only a charge +e to the unit cell, compared to −2e from the phosphates. As will be seen, however, the hole wavefunction is not localized on one base but spread over a number of them.

We address first the modification of the Hamiltonian to include the interaction with the environment. This may be carried out by adding a term to the Hamiltonian representing the change in the hole energy due to the change in polarization it causes. Taking the hole charge density as $\rho(r)$, we may write the energy of interaction of the hole with the environment

$$E_{\mathrm{env}} = (1/2)\int d^3r d^3r' \rho(r)[G(r,r') - (1/|r-r'|)]\rho(r'), \tag{7}$$

where $G(r,r')$ is the electrostatic Green's function in this environment, given by

$$\varphi(r) = \int d^3r' G(r,r')\rho(r'), \tag{8}$$

with $\phi(r)$ being the electrostatic potential. Equation 7 represents the electrostatic energy of the charge in the presence of the environment, with the energy of interaction of the charge with itself subtracted.

We represent the hole wavefunction as a linear combination of molecular orbitals: $\psi(r)=\Sigma_n\psi_n\phi_n(r-r_n)$, where ϕ_n is the orbital of the nth base and ψ_n the probability amplitude for the hole on this base. Neglecting the orbital overlap for different bases, we obtain the corresponding charge density

$$\rho(r) = e|\psi(r)|^2 \approx e\sum_n |\psi_n|^2 \phi_n^2(r-r_n) \tag{9}$$

Inserting Eq. 9 into Eq. 7, and denoting the integrals over r,r' for each pair of terms in the sum Eq. 9 by $g_{n-n'}$, we express the hole energy as

$$H(\psi_n,\psi_n^*) = H_0(\psi_n,\psi_n^*) + (1/2)\sum_{n,n'} g_{n-n'}|\psi_n|^2|\psi_{n'}|^2. \tag{10}$$

We consider first the case where H_0 corresponds to free hole motion, given by Eq. 2 with α=0 and also A=0 because we are discussing only static solutions here. In all that follows we restrict ourselves to the case where the DNA stack consists of the same base pair repeated, which means that d_n in Eq. 2 may also be taken equal to zero.

A practical way of describing the environment is to consider the DNA molecule to be placed inside a cavity. Physically, the cavity is due to the sugar-phosphate backbone and the hydrophobicity of the DNA bases. Its characteristic size R is determined by the radius of the helix: R~10 Å. The space outside the cavity is filled with water, with dielectric constant 78 at room temperature, in which counterions are dissolved with equilibrium densities (far from the DNA molecule) $n_{\mathrm{Na}}=n_{\mathrm{Cl}}$. Neglecting the polarizability of the backbone and the bases, we assume the cavity to be empty. It is readily shown that, because of the large dielectric constant of water, the interaction with the ions may be neglected [64].

To calculate the interaction with water it is reasonable to consider a cylindrical cavity with radius R and to assume the charge to be concentrated on the axis of the cylinder at evenly spaced points $z=na$ corresponding to the bases. Taking advantage of the cylindrical symmetry, we make the Fourier transform

$$\rho(r) = \delta(x)\delta(y) \int dk(\rho_k/2\pi)e^{ikz}, \text{where integration is from} - \infty \text{ to } \infty. \quad (11)$$

The corresponding Fourier component of the potential $\phi_k(r_\perp)$, where $r_\perp \equiv \sqrt{(x^2+y^2)}$, is found by solving Laplace's equation with the appropriate boundary conditions [64]. With this solution we find the interaction energy

$$E_{\text{env}} = -\int dk[K_0(|k|R)/I_0(|k|R)]\rho_k\rho_{-k}/2\pi, \text{where integration is from } -\infty \text{ to } \infty, \quad (12)$$

K_0 and I_0 being modified Bessel functions. The inverse Fourier transform and comparison with Eq. 10 lead to

$$g_n = -(e^2/R)v(na/R), \quad (13)$$

where

$$v(\xi) = (2/\pi) \int [K_0(q)/I_0(q)] \cos q\xi dq, \text{where integration is from 0 to } \infty \quad (14)$$

Setting $R=3a=10.2$ Å we may calculate the g values from the last two equations. We find that $g_0=-0.87\ e^2/R$, close to the value obtained for a spherical cavity [64], testifying to the reliability of this method of estimation. We find also that g_n varies slowly with n, due to the fact that $a<R$. For t_0~0.1–0.2 eV this means that the kinetic energy will not allow the hole to be localized on a single site. Balancing the kinetic and interaction terms requires that the hole be spread over three to five sites. The significance of this result is that the shape of the hole wavefunction will not be strongly different from that obtained without taking into account solvation effects. Significantly, however, the solvation increases the binding energy by ~$|g_0|/2$, or ~0.6 eV.

That the wavefunction is not much changed by solvation is seen in Fig. 4, where the hole population profile $|\psi_n|^2$ is shown for coupling to the lattice displacements and the environment separately and together. The parameters used are t_0=0.2 eV, α^2/K=0.2 eV. The lattice contribution can be included in the hole Hamiltonian (Eq. 10) by adding $-2\alpha^2/K$ to $g_{\pm 1}$ from Eqs. 13 and 14.

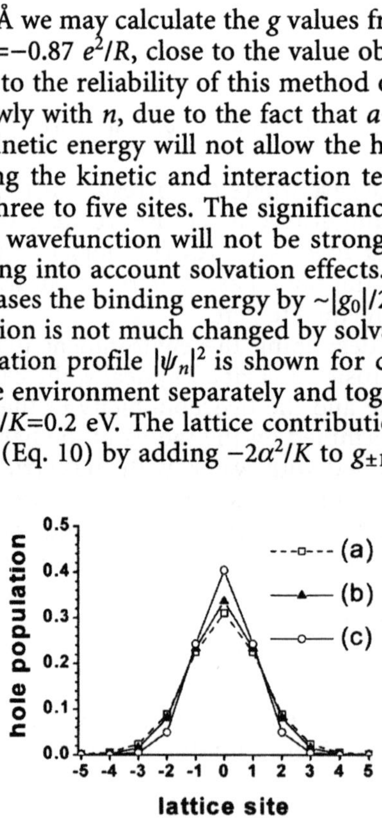

Fig. 4a–c Hole population, $|\psi_n|^2$, for a stationary state formed by interaction with **a** lattice deformation only, **b** environment only, **c** both (After Basko and Conwell [64])

The binding energies corresponding to the curves (a), (b), and (c) are 0.057 eV (lattice only), 0.52 eV (environment only), and 0.62 eV (both), respectively. The shapes are quite similar to each other and are weakly dependent on the binding energy. For t_0=0.1 eV the population profile due to interaction with the environment only is indistinguishable from the curve (c) and the binding energy is 0.56 eV. From these results we conclude that the stationary state of DNA in the physiological environment is polaronic in nature at room temperature and for a wide range of temperatures above and below it.

For a DNA molecule in air or vacuum it is much more difficult to calculate the contribution of the environment to the binding energy of a polaron. When the DNA is taken out of solution, positive charge—protons, Na ions, other positive ions in the air—will collect on the surface to neutralize the negative charge of the backbone. The surface charge provides a polarizable medium that will act to lower the energy of an excess charge on the base stack. It is clear that the effect will be less than that of water surrounding the DNA, but in the absence of knowledge about the nature and distribution of the condensed charge it is not possible to quantify this effect. The situation is made even more difficult in that DNA in air is not in B form unless the relative humidity is kept very high.

2.5 Properties of Moving Polarons

From the Hamiltonian Eqs. 1–3 we derive the equations of motion of the polaron:

$$i\hbar(\partial\psi_n/\partial t) = d_n\psi_n - [t_0 - \alpha(u_{n+1} - u_n)]e^{-i\gamma A}\psi_{n+1} - [t_0 - \alpha(u_n - u_{n-1})]e^{i\gamma A}\psi_{n-1} \quad (15)$$

$$M(\partial^2 u_n/\partial t^2) = K(u_{n+1} - 2u_n + u_{n-1}) + \alpha\left[e^{i\gamma A}\left(\psi_n^*\psi_{n+1} - \psi_{n-1}^*\psi_n + c.c.\right)\right] \quad (16)$$

These equations do not include the effect of the surrounding medium; that will be taken up later in this section.

We use these equations first to study formation of the polaron in the absence of an electric field, i.e., with A=0 [65]. Because only small motions of the sites are involved in the formation, the results should be approximately correct even for DNA in solution. To begin with we find the static undistorted solution (y_n=0) for a stack of N base pairs with $2N$ π electrons. Using this solution as the initial condition for t=0, we integrate Eqs. 15 and 16 numerically for the case of $2N-1$ π electrons on the stack. The parameters used were t_0=0.3 eV, α=0.6 eV/Å, and K=0.85 eV/Å^2. The result for the sequence given at the bottom of the figure is shown in Fig. 5. It is seen that the polaron is fully formed at 4 ps. The time of formation is much longer than was found for polarons in polyacetylene. The calculations of Su and Schrieffer

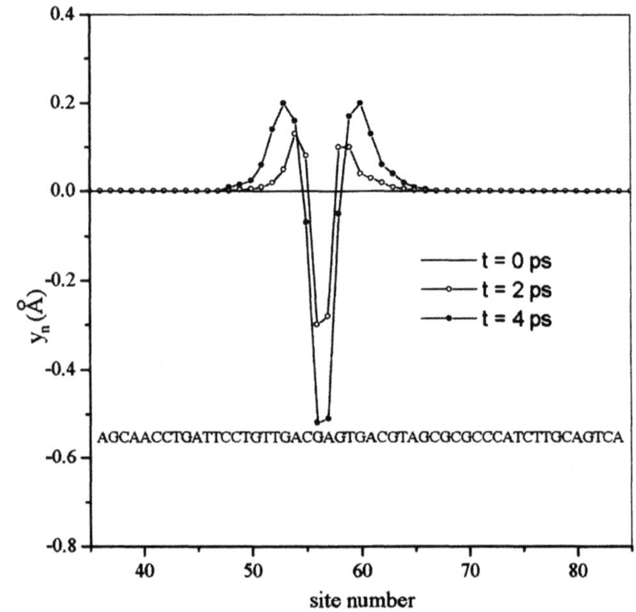

Fig. 5 Polaron configuration vs. time, i.e., polaron formation, for zero field (From Rakhmanova and Conwell [65])

gave the time for polaron formation in polyacetylene as only tenths of a picosecond [66], in agreement with later experiments on that material. The longer formation time is the result of the sites being much more massive and the coupling to acoustic modes smaller.

To investigate polaron motion we started on the stack, at t=0, a polaron obtained by solution of Eqs. 1–5 with A=0. To apply a constant electric field we took $A=A_0t$. The field is then $-A_0/c$. A_0 was chosen to give a moderate field, 5×10^3 V/cm. For this field numerical integration of the equations of motion gave a polaron drifting smoothly, maintaining its shape, when the stack consisted of the same base pair repeated (Fig. 6).

When there was a random sequence of base pairs the polaron moved only a few bases at the most, the large differences in site energies functioning as barriers. That is, as was discussed earlier, in agreement with experiment. In the stack with the same base pair repeated the polaron was found to move 15.6 lattice sites in a field of 5.8×10^3 V/cm. Because no scattering was included in the equations of motion, nor any effect due to the environment, the polaron undergoes constant acceleration of $e\hat{E}/m_p$, where $\hat{E}$ is the field strength and m_p is the mass of the polaron. Given the distance traveled and the time required for that distance, we can obtain the acceleration and calculate the mass of the polaron. The result is that the mass of the polaron is 0.05 times the mass of a C:G or A:T pair.

The motion described in the last paragraph, although it might be approximately correct for a polaron on DNA in air or vacuum, is not applicable for

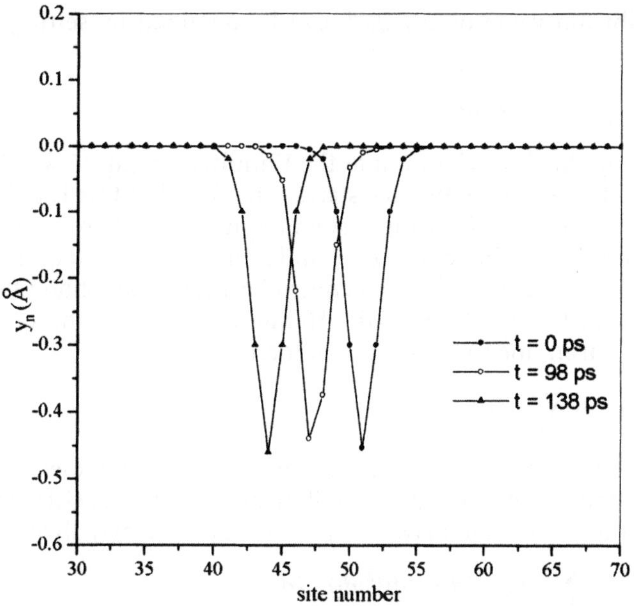

Fig. 6 Motion of a polaron in DNA made up of a single base pair repeated, in a field of 5.8×10^3 V/cm, not including the effects of the solution (From Rakhmanova and Conwell [65])

DNA in solution. The drag of the water on the polaron in motion results in its achieving a steady state with a terminal drift velocity v_d. The terminal velocity is obtained when the rate at which the polaron gains energy from the field equals the rate at which it loses energy to the water. Having determined the interaction of the polaron with water in the last section, we can find the mobility corresponding to the terminal drift velocity. In detail, the moving polaron causes a reorientation of the water dipoles, inducing a screening charge to be denoted Q_n. The effect of the polaron on the ions in solution also creates a screening charge (Debye–Hückel). The latter charge can be neglected, however, because the characteristic time τ_i, related to the conductivity σ and dielectric constant ε by $\tau_i=\varepsilon/4\pi\sigma$, is 650 ps at room temperature for the 0.1 M solution of NaCl, whereas the orientational relaxation time for the water molecules is 8.3 ps.

We now set up a Hamiltonian for the polaron interacting with the screening charge due to the water [64]:

$$H(\psi_n,\psi_n^*,Q_n)=H_0(\psi_n,\psi_n^*)+\sum_{n,n'}\lambda_{n-n'}Q_n|\psi_{n'}|^2+(\kappa/2)\sum_n Q_n^2 \qquad (17)$$

where H_0 is, as in Eq. 10, the Hamiltonian for free polaron motion (Eq. 2 with d_n, α, and A set equal to zero). The term representing the coupling between the polaron and the screening charge is linear in Q_n and $|\psi_n'|^2$ because both of these represent charges and the coupling is of Coulomb origin. The last term represents the energy contribution of the screening charge. The static equilibrium value of Q_n, Q_n^{eq}, can be obtained by setting $\partial H/\partial Q_n=0$. This gives

$$Q_n^{eq} = -\sum_{n'} (\lambda_{n-n'}/\kappa)|\psi_{n'}|^2 \tag{18}$$

Substituting this last relation into the Hamiltonian Eq. 17 we obtain an effective Hamiltonian for stationary states. The fact that this Hamiltonian and that of Eq. 10 represent the same system requires that there be a relation between $g_{n-n'}$ on the one hand and $\lambda_{n-n'}$ and κ on the other. Requiring also that $\Sigma Q_n^{eq}=-\Sigma|\psi_n|^2$, because Q_n is a screening charge, we can obtain this relation.

According to the last term of the Hamiltonian Eq. 17 the rate of energy loss of the polaron due to the frictional force of the water is

$$(dE/dt) = -\tau\kappa \sum_n (dQ_n/dt)^2 \tag{19}$$

where τ is the orientational relaxation time of a water molecule. If the polaron is moving slowly then Q_n will follow quasistatically: $Q_n(t)\approx Q_n^{eq}(t)$. Using Eq. 18 and the relation between $\lambda_{n-n'}$ and $g_{n-n'}$ we may then write:

$$(dE/dt) = -\tau \sum_n g_{n-n'} |d\psi_n/dt|^2 |d\psi_{n'}/dt|^2 \tag{20}$$

Because the polaron maintains its shape while moving we may take $|\psi_n|^2=F(n-v_d t/a)$, with a being the lattice constant, 3.4 Å, and $F(\xi)$ being a continuous function. We may then take $d\psi_n/dt=F'(n)$, where $F'(\xi)=dF/d\xi$. Using the definition of mobility $\mu\equiv v_d/\hat{E}$, where $\hat{E}$ is the electric field strength, we may also write $dE/dt=ev_d^2/\mu$. Equating this form of (dE/dt) with Eq. 20 we obtain

$$(e/\mu) = -(\tau/a^2)\sum_{n,n'} g_{n-n'}F'(n)F'(n') \tag{21}$$

To evaluate the sum we take $F(n)=|\psi_n|^2$ for the standing polaron, namely, curve (c) in Fig. 4, and define the derivative of the lattice function $F(n)=|\psi_n|^2$ through the Fourier transform $[F'(n,k)\leftrightarrow ikF(n,k)]$. The result is $\mu=2.4\times10^{-3}$ cm^2/Vs for the limit of small drift velocity. From μ, using the Einstein relation, we obtain an upper limit for the diffusion constant D, which could be measured experimentally. Our calculations give the upper limit of D for a hole diffusing on a DNA stack with all base pairs the same as 3×10^{-5} cm^2/s. In the opposite limit of high drift velocity, the solvent will not follow the hole and the dissipation will be negligible. The high-field mobility and diffusion constant will then be determined by different effects, which will not be considered here.

3
Doping of DNA

When the relatively low conductivity of DNA was acknowledged, there were suggestions that DNA be doped to increase its conductivity. It is immediately apparent that the methods used for doping conjugated polymers, which led in favorable cases to increases in conductivity of factors of 10^{10} and more, might not be successful for DNA. Na^+ ions are good dopants for the conjugated polymers. Each Na goes between the polymer chains and donates a free electron to a chain. But DNA in a solution of Na^+ ions does not show evidence of free charges. In fact it has been suggested that an Na^+ ion caught in the major groove might result in a trapped electron on the base stack [63].

Nevertheless, efforts to dope DNA films in air with oxygen have met with some success [67]. Oxygen acts as an acceptor and has been found to increase hole concentration in conducting polymers. The films to be doped were made by dropping a DNA sample diluted with deionized water on Au/Ti electrodes separated by ~100 μm, and subsequently drying in air. The work was done with two kinds of film: poly(dA)-poly(dT) and poly(dG)-poly(dC), each with the constituent DNA molecules ~1 μm in length. The former composition is expected to result in an n-type film, the latter p-type [27]. These expectations were verified by Hall measurements on the films. On exposure to oxygen after having been in a vacuum chamber, films of poly(dG)-poly(dC) were found to have increased conductance, as expected for hole generation in p-type material, by as much as a few orders of magnitude. Exposure to oxygen of poly(dA)-poly(dT) films resulted in a decrease in conductance, as expected for n-type films.

The effect of humidity on conductivity of thin films of poly(dG)-poly(dC) and poly(dA)-poly(dT) in air was also investigated. It was found that above 20% relative humidity (RH) the conductance increases exponentially with RH [68]. It was concluded that the resistance of the films is not determined by that of the DNA, but rather by conduction through the hydration layers surrounding the DNA molecules in the film [68].

Doping that took into account the chemical structure of DNA was reported by Rakitin et al. [26]. The doping process consists of substituting the imino proton of each base pair with a metal ion, usually Zn^{++}. It was carried out by adding to B-DNA (consisting of λ-DNA ~15 μm long) 0.1 mM Zn^{++} at pH 9.0 [69]. The resulting DNA was called M-DNA. Although the claim was made that M-DNA is metallic, comparison of current I vs voltage V in vacuum of M-DNA with that of B-DNA showed them to be essentially identical at negative voltages [26]. For positive voltages the I–V curves of the two differed because there was a threshold for current flow of ~0.2 V for the B-DNA, although no threshold for the M-DNA. Beyond 0.2 V the B-DNA I–V curve was parallel to that of the M-DNA [26]. Crudely, the resistance of the M-DNA is about one tenth of that seen by Porath et al. [28] but this DNA is ~100 times as long. However, the DNA of these experiments is a "rope" with thickness estimated as ~3×10^2 times that of a single duplex strand.

In any case, despite the claim of [26], M-DNA does not show metallic conduction. What the doping has accomplished is to bring down the threshold voltage required for the positive direction of voltage from 0.2 V to zero. Since the likely explanation for the existence of a threshold voltage is a contact barrier arising from mismatch of the Fermi energy in the metal with that in the DNA [28], the doping seems to have eliminated the barrier at the contact. Rakitin et al. [26] speculate that this is a result of the metal ion insertion creating a d-band aligned with the electrode Fermi level.

Finally, it is always possible to make DNA a conducting wire by coating it with silver or other metal, as done by Braun et al. [8]. The trick would be to do this while retaining the self-assembly properties of DNA. Alternatively, one could make use of the self-assembly properties before carrying out the coating, as done by Braun et al.

4
Comments on Conductivity Calculations

Among the theoretical calculations that have been carried out for transport in DNA made up of a series of identical base pairs are those where the wavefunction is assumed to be localized on a single base or base pair, and those that assume completely delocalized, or free, electrons and holes. In the former case the confinement to one base is, at least implicitly, assumed due to the transfer integral being quite small rather than to traps in the DNA. Calculations that assume the effect of the transfer integral is nullified by strong thermal vibrations (which are, nevertheless, responsible for the interbase motion of the carriers) were carried out by Bruinsma et al. [34], discussed in Sect. 1.2, and Smith and Adamowicz [70]. Microscopic calculations of hopping transport between adjacent bases, based on electron transfer theory, were carried out by Jortner et al. [9]. These calculations will not be discussed here; a large part of the foregoing material in this chapter, particularly the section on "Why Large Polarons?", has been devoted to making the case that localization to one site does not take place in defect-free DNA with all the same base pairs.

The calculations assuming free electrons and holes are for the most part intended to apply to the case of DNA in air or vacuum. The binding energy of the polaron calculated in Sect. 2.1, without any contribution from the solvent, is only a few kT at room temperature. As discussed in Sect. 2.4, the binding energy in air or vacuum is likely to be enhanced by the surface layer of positive ions required to neutralize the negative charge of the backbone, but the enhancement will not be as great as was found in that section for the polaron in solution. Thus the stability of the polaron is not certain at room temperature, although it should certainly be stable at lower temperatures. An argument that the carriers are polarons through the entire temperature range is the fact that the current–voltage characteristics found by Porath et al. [28] showed little dependence on temperature from 4 to 300 K other than the increase in the size of the voltage gap.

A key question about the transport in air or vacuum is how disordered is the DNA, particularly due to the relative absence of water. It is difficult to believe, as assumed in the calculation of Hjort and Stafstrom [71], that the DNA structure would be the same as that of a similar sequence in a crystal determined by x-rays. Note also that disorder, in being unfavorable for free carriers, is favorable for polarons.

The mobility estimates of Grozema et al. [52] are quite optimistic, at least so far as hole transport in solution is concerned. For what are called "realistic rotational force constants and static energy disorder" they predict a hole mobility in DNA of 0.1 cm^2/Vs. However, taking into account the drag effect of the water we obtained a mobility of 3.5×10^{-3} cm^2/Vs, as detailed in Sect. 2.5. This is an upper limit because we included neither scattering nor energetic disorder.

5 Concluding Remarks

We have calculated the wavefunction of an extra hole or electron on the base stack of DNA using a simple tight-binding model. In agreement with theoretical predictions for a 1D case we find the wavefunction is not localized on a single base or base pair. Both in air and in solution, for reasonable values of the parameters, we find the wavefunction extended over five or six sites independent of the base sequence, although the detailed shape does depend on sequence. Good evidence for extended wavefunctions is the finding that calculations based on the tight-binding Hamiltonian that gives polarons lead to excellent agreement with experiment for the depths of GG and GGG traps relative to G. This agreement is very little dependent on the value of the transfer integral. By contrast, the model where the wavefunction of the extra hole is limited to one site leads to poor agreement with experiment. Other arguments for extended wavefunctions that last a minimum of nanoseconds are based on what is known about excitons and excimers in DNA.

When the sequence over which the polaron extends includes one or more Gs the stationary wavefunction is centered on the Gs, as is seen in Figs. 1–3. Other bases have lower HOMO levels than G, thus constituting barriers to polaron motion. The experimental findings of Giese's group on hole propagation in a sequence where Gs are separated by up to several A:Ts can be reinterpreted as showing hole polaron tunneling between different potential wells, the wells defined by the location and number of the Gs. With this model we can account for the switch from tunneling to on-bridge propagation as the number of A:Ts between the Gs goes beyond three. The rapid, almost unattenuated, hole motion found by Giese's group after thermal energy allows the hole to move onto the bridge of As is seen as polaron drift rather than incoherent hole hopping between As.

We have also calculated the effect on the polaron of the water and counterions surrounding the DNA under physiological conditions. Little change was found in the extent of the polaron, but the binding energy was greatly increased. The drag due to the water results in limiting values for the mobility and diffusion constant, which were evaluated.

References

1. Eley DD, Spivey DI (1962) Trans Faraday Soc 58:411
2. Magan JD, Blau W, Croke DT, McConnell DJ, Kelly JM (1987) Chem Phys Lett 141:489
3. Spalletta RA, Bernhard WA (1992) Rad Res 130:7
4. Razskazovskii Yu, Swarts S, Falcone J, Taylor C, Sevilla M (1997) J Phys Chem 101:1460
5. Tran P, Alavi B, Gruner G (2000) Phys Rev Lett 85:1564
6. Holmlin R, Dandliker P, Barton J (1997) Angew Chem Int Ed 36:2714
7. Burrows CJ, Muller JG (1998) Chem Rev 98:1109
8. Braun E, Eichen Y, Sivan U, Ben-Yoseph G (1998) Nature 391:775
9. Jortner J, Bixon M, Langenbacher T, Michel-Beyerle, M (1998) Proc Natl Acad Sci USA 95:12759
10. Meggers E, Michel-Beyerle M, Giese B (1998) J Am Chem Soc 120:12950
11. Giese B (2000) Acc Chem Res 33:631
12. Schuster GB (2000) Acc Chem Res 33:253
13. Giese B, Amaudrut J, Kohler, A-K, Spormann M, Wessely S (2001) Nature 412:318
14. Hush NS, Cheung AS (1975) Phys Lett 34:11
15. Lifschitz C, Bergman E, Pullman B (1967) Tetrahedron Lett 46:4583
16. Seidel CAM, Schultz A, Sauer MHM (1996) J Phys Chem 100:5541
17. Steenken S, Jovanovic SC (1997) J Am Chem Soc 119:617
18. Bixon M, Jortner J (2001) J Am Chem Soc 123:12556
19. Kendrick T, Giese B (2002) Chem Comm 18:2016
20. Giese B, Wessely S, Spormann M, Lindemann U, Meggers E, Michel-Beyerle ME (1999) Angew Chem Int Ed 38:996
21. Behrens C, Burgdorf LT, Schwogler A, Carell T (2002) Angew Chem Int Ed 41:1763
22. de Pablo PJ, Moreno-Herrero F, Colchero J, Gomez Herrero, J, Herrero P, Baro AM, Ordejon P, Soler JM, Artacho E (2000) Phys Rev Lett 85:4992
23. Gomez-Navarro C, Moreno-Herrero F, de Pablo PJ, Colchero J, Gomez-Herrero J, Baro AM (2002) Proc Natl Acad Sci USA 99:8484
24. Kasumov AYu, Kociak M, Gueron S, Reulet B, Volkov VT, Klinov DV, Bouchiat H (2001) Science 291:280
25. Fink H-W, Schonenberger C (1999) Nature 398:407
26. Rakitin A, Aich P, Papadopoulos C, Kobzar Yu, Vedeneev AS, Lee JS, Xu JM (2001) Phys Rev Lett 86:3670
27. Yoo K-H, Ha DH, Kee J-O, Park JW, Kim J, Kim JJ, Lee H-Y, Kawai T, Choi H-Y (2001) Phys Rev Lett 87:198102
28. Porath D, Bezryadin A, de Vries S, Dekker C (2000) Nature 403:635
29. Bottger H, Bryksin VV (1985) Hopping conduction in solids. Akademie, Berlin
30. Basko DM, Conwell EM (2002) Phys Rev B 66:094304
31. Emin D, Holstein T (1976) Phys Rev Lett 36:323
32. Davis WB, Svec WA, Ratner MA, Wasielewski MR (1998) Nature 396:60
33. Conwell EM (1997) Transport in conducting polymers In: Nalwa HS (ed) Handbook of organic conductive molecules and polymers, vol 4. Wiley, New York, p 1
34. Bruinsma R, Gruner G, D'Orsogna MR, Rudnick J (2000) Phys Rev Lett 85:4393
35. Rist M, Wagenknecht H-A, Fiebig T (2002) Chemphyschem 3:704
36. Wan C, Fiebig T, Kelley SO, Treadway CR, Barton JK, Zewail AH (1999) Proc Natl Acad Sci USA 96:6014
37. Lewis FJ, Wu T, Liu X, Letsinger RL, Greenfield SR, Miller SE, Wasielewski MR (2000) J Am Chem Soc 122:2889
38. Brauns EB, Madaras ML, Coleman RS, Murphy CJ, Berg MA (1999) J Am Chem Soc 121:11644
39. Swaminathan S, Ravishankar G, Beveridge D (1991) J Am Chem Soc 113:5027
40. Bredas JL, Themans B, Andre JM (1982) Phys Rev B 26:6000

41. Michl J, Bonacic-Koutecky V (1990) Electronic aspects of photochemistry. Wiley, New York
42. Gueron M, Eisinger B, Lamola AA (1974) Excited states of nucleic acids. In: Ts'o POP (ed) Basic principles in nucleic acid chemistry, vol 1. Academic, New York, p 312
43. Ge G, Georghiou S (1991) Photochem Photobiol 54:301
44. Rigler R, Claessens F, Kristensen O (1985) Anal Instrum 14:525
45. Su WP, Schrieffer JR, Heeger AJ (1980) Phys Rev B 22:2099
46. Conwell EM, Rakhmanova SV (2000) Proc Natl Acad Sci USA 97:4556
47. Sugiyama H, Saito I (1996) J Am Chem Soc 118:7063
48. Conwell EM, Basko DM (2001) J Am Chem Soc 123:11441
49. Voityuk AA, Rosch N, Bixon M, Jortner J (2000) J Phys Chem B 104:9740
50. Voityuk AA, Siriwong K, Rosch N (2001) Phys Chem Chem Phys 3:5421
51. Troisi A, Orlandi G (2001) Chem Phys Lett 344:509
52. Grozema FC, Siebbeles LDA, Berlin YA, Ratner MA (2002) Chemphyschem 3:536
53. Hickerson RP, Prat F, Muller JG, Foote CS, Burrows CJ (1999) J Am Chem Soc 121:9423
54. Berlin YA, Burin AL, Ratner MA (2001) J Am Chem Soc 123:260
55. Voityuk AA, Jortner J, Bixon M, Rosch N (2001) J Chem Phys 114:5614
56. Lewis FD, Liu X, Hayes RT, Wasielewski MR (2000) J Am Chem Soc 122:12037
57. Voityuk AA, Jortner J, Bixon M, Rosch N (2000) Chem Phys Lett 324:430
58. Spassky A, Angelov D (1997) Biochemistry 36:6571
59. Sistare MF, Codden SJ, Heimlich G, Thorp HH (2000) J Am Chem Soc 122:4742
60. Basko DM, Conwell EM (2002) Phys Rev E65:061902
61. Bixon M, Giese B, Wessely S, Langenbacher T, Michel-Beyerle ME, Jortner J (1999) Proc Natl Acad Sci USA 96:11713
62. Manning GS (1969) J Chem Phys 51:924
63. Barnett RN, Cleveland CL, Joy A, Landman U, Schuster GB (2001) Science 294:567
64. Basko DM, Conwell EM (2002) Phys Rev Lett 88:098102
65. Rakhmanova SV, Conwell EM (2001) J Phys Chem B 105:2056
66. Su WP, Schrieffer JR (1980) Proc Natl Acad Sci USA 77:5626
67. Lee H-Y, Tanaka H, Otsuka Y, Yoo K-H, Lee J-O, Kawai T (2002) Appl Phys Lett 80:1670
68. Ha DH, Nham H, Yoo K-H, So H-M, Lee H-Y, Kawai T (2002) Chem Phys Lett 355:405
69. Aich P, Labiuk SL, Tari L, Delbaere JT, Roesler WJ, Falk KJ, Steer RP, Lee JS (1999) J Mol Biol 294:477
70. Smith DMA, Adamowicz L (2001) J Phys Chem B 105:9345
71. Hjort M, Stafstrom S (2001) Phys Rev Lett 87:228101

Top Curr Chem (2004) 237:103–127
DOI 10.1007/b94474

Studies of Excess Electron and Hole Transfer in DNA at Low Temperatures

Zhongli Cai · Michael D. Sevilla

Department of Chemistry, Oakland University, Rochester, MI 48309, USA
E-mail: sevilla@oakland.edu

Abstract In this review investigations of hole and electron transfer processes in DNA after γ-irradiation at low temperatures are described. These experiments suggest that at low temperatures DNA is a good ion radical trap, one which traps 30 to 60% of all electrons and holes produced by radiation. Electrons are trapped at the pyrimidine bases, cytosine and thymine, whereas holes are trapped mainly on guanine. Results show that all electrons in the hydration layer transfer to DNA and all holes in the first layer of waters transfer to DNA; holes in subsequent layers form hydroxyl radicals. After trapping, electron migration and hole transfer processes within DNA are limited to tunneling at low temperatures. Electron spin resonance studies have followed the time-dependent transport of electrons and holes from DNA base trap sites to acceptors such as intercalators or modified bases. Electron transfer through the DNA π stack and between duplexes has been elucidated, as well as the effects of DNA hydration, complexing agents, base sequence, and H/D isotope exchange on electron-transfer distances and rates. Studies which vary the temperature have separated tunneling in DNA from activated mechanisms that dominate at temperatures near 200 K and above.

Keywords Tunneling · Radiation · Low temperature · Electron spin resonance · DNA

Abbreviations

ESR	Electron spin resonance
ET	Electron transfer
C	Cytosine
T	Thymine
G	Guanine
A	Adenine
MX	Mitoxantrone
ν	Mole ratio of intercalator or modified base to DNA base pairs
bp	Base pair
β	Tunneling constant
EtBr	Ethidium bromide
Pa	1,10-Phenanthroline
NPa	5-Nitro-1,10-phenanthroline
EA	Electron affinity
ds	Double strand
D_a	Apparent transfer distance
D_I	Actual transfer distance along the primary DNA ds
D_{ds}	Interduplex center-to-center separation
SP	Spermine tetrahydrochloride
DOD	Dodecyltrimethylammonium bromide
OCT	Octadecyltrimethylammonium bromide
PLL	Poly-L-lysine hydrobromide
PEI	Polyethylenimine hydrochloride
IE	Ionization energy
polydIdC-polydIdC	Polydeoxyinosinic-deoxycytidylic acid sodium salt
polydAdT-polydAdT	Polydeoxyadenylic-thymidylic acid sodium salt
polydGdC-polydGdC	Polydeoxyguanylic-deoxycytidylic acid sodium salt
polydG-polydC	Polydeoxyguanylic-polydeoxycytidylic acid sodium salt
polyC-polyG	Polycytidylic-polyguanylic acid sodium salt
polyA-polyU	Polyadenylic-polyuridylic acid sodium salt
8-oxo-G	8-Oxo-7,8-dihydroguanine
G_{ox}	Further oxidation products of 8-oxo-G

Γ	Number of water molecules per nucleotide
T(OH)Br	5-bromo-6-hydroxy-5,6-dihydrothymine
CBr	5-bromocytosine
GBr	5-bromoguanine

1 Introduction

The rate and extent of electron and hole migration within DNA had been a topic of intense experimental [1] and theoretical [2] interest and dispute. However, as other chapters in this volume point out, the overall picture is now becoming increasingly understood. Several earlier reviews have dealt with the radiation damage to DNA and subsequent chemical processes including electron and hole transfer [3]. This chapter will focus on recent contributions to hole and electron transfer in DNA from high-energy radiation studies. We first present a brief overview on electron transfer processes in DNA from radiation studies and describe in more detail our most recent results on electron and hole transfer in DNA at low temperatures.

2 Brief Review of Studies of Electron and Hole Formation and Transfer After Direct Ionization of DNA

2.1 State of Our Understanding on Radiation-Induced Free Radical Formation in DNA

Direct ionization of DNA produces electrons and holes randomly throughout the DNA structure and its intimate surroundings roughly in proportion to the number of valence electrons at each site. Hole transfer from these initial sites of ionization is rapid as these are most often not at sites of lowest potential energy. Ionization-produced electrons are ejected with varying energies, thermalize in a few nanometers, and are trapped at sites of high electron affinity. About 30 to 60% of all initial radiation-produced ion radicals are trapped in DNA at low temperatures depending on DNA structure and hydration level [4, 5]. Electron spin resonance (ESR) results at 77 and 4 K suggest that the average trapped electron-to-hole distance is 3 to 5 nm at low temperatures [4–6]. Recombination of the anion and cation radical species occurs via tunneling so that, on the time frame of minutes, all trapped electrons and holes within 3 nm recombine. Over longer times further electron and hole transfer via tunneling occurs to deeper traps that extends this distance outward [7] and is discussed in detail below. On warming DNA samples to higher temperatures the trapped ions overcome barriers toward hopping [1i, 8]. However, this hopping likely involves proton-coupled electron transfer (PCET). Subsequent hops are either hindered by reversible or terminated by irreversible protonation reactions for electron adducts and deprotonation reactions for holes [9–12].

2.2
Electron Localization at Low Temperatures

ESR experiments at low temperatures with frozen aqueous solutions of DNA indicate electrons are trapped on cytosine (C, 50–85%) with the remainder on thymine (T) [6, 9,13]. After the initial electron attachment, electron migration is temporarily quenched at low temperatures. The reversible protonation of the cytosine anion, $C^{\cdot-}$ at N3 to form the neutral species, $C(N3)H^{\cdot}$ is suggested to further stabilize this species (Scheme 1). Much experimental

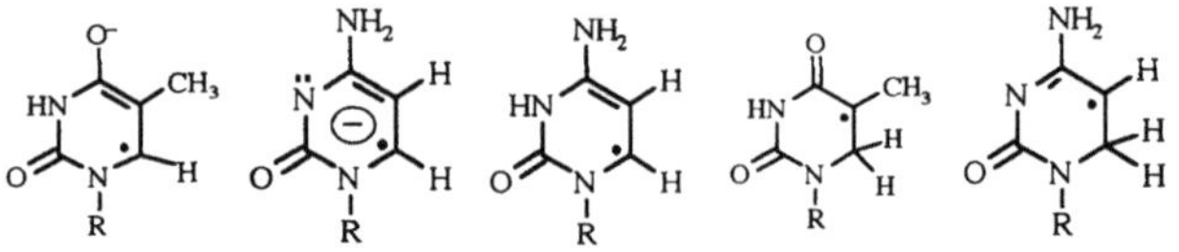

Scheme 1 Chemical structure of main electron localization species of DNA at low temperatures. In D_2O solutions $C(N3)H^{\cdot}$ is described as $C(N3)D^{\cdot}$ or $CD^{\cdot}$

[11, 12] and theoretical [14–16] work confirms this proton transfer process and its importance to any subsequent excess electron transport. Figure 1 shows a ball-and-stick model of the GC anion radical, giving the lengths of hydrogen bonds during proton transfer (Fig. 1a), as well as energy (uncorrected for ZPE) profiles as the guanine (G) N1-H distance increases (Fig. 1b) [14]. While at low temperatures electron transfer is confined to slow long-range tunneling [7], on annealing to higher temperatures (ca. 190 K) more rapid electron migration via hopping ensues. This transfer is finally irreversibly quenched by the protonation of the thymine anion and the resulting formation of the 5,6-dihydrothymin-5-yl radical, $T(C6)H^{\cdot}$ [9, 10]. Since $T(C6)H^{\cdot}$ formation accounts for the majority of the electron-gain centers, and the electron is largely localized at cytosine at low temperatures, it is clear that transfer of the electron from cytosine to thymine occurs [9].

2.3
Temperature and Proton Transfer Effects on Electron Transfer

In order to understand the effect of temperature we consider the energetics involved for the stepwise formation of $T(C6)H^{\cdot}$ in DNA. The proton-coupled electron transfer from cytosine to thymine in DNA is depicted in reaction 1 and protonation of the thymine anion radical in reaction 2.

$$\mathrm{DNA}[\mathrm{C(N3)H}^{\cdot},\mathrm{T}] \leftrightarrows k_1 k_{-1} \mathrm{DNA}[\mathrm{C},\mathrm{T}^{\cdot-}] \quad (1)$$

$$\mathrm{DNA}[\mathrm{T}^{\cdot-}] \leftrightarrows k_2 + \mathrm{H}^+ \mathrm{DNA}[\mathrm{T(C6)H}^{\cdot}] \quad (2)$$

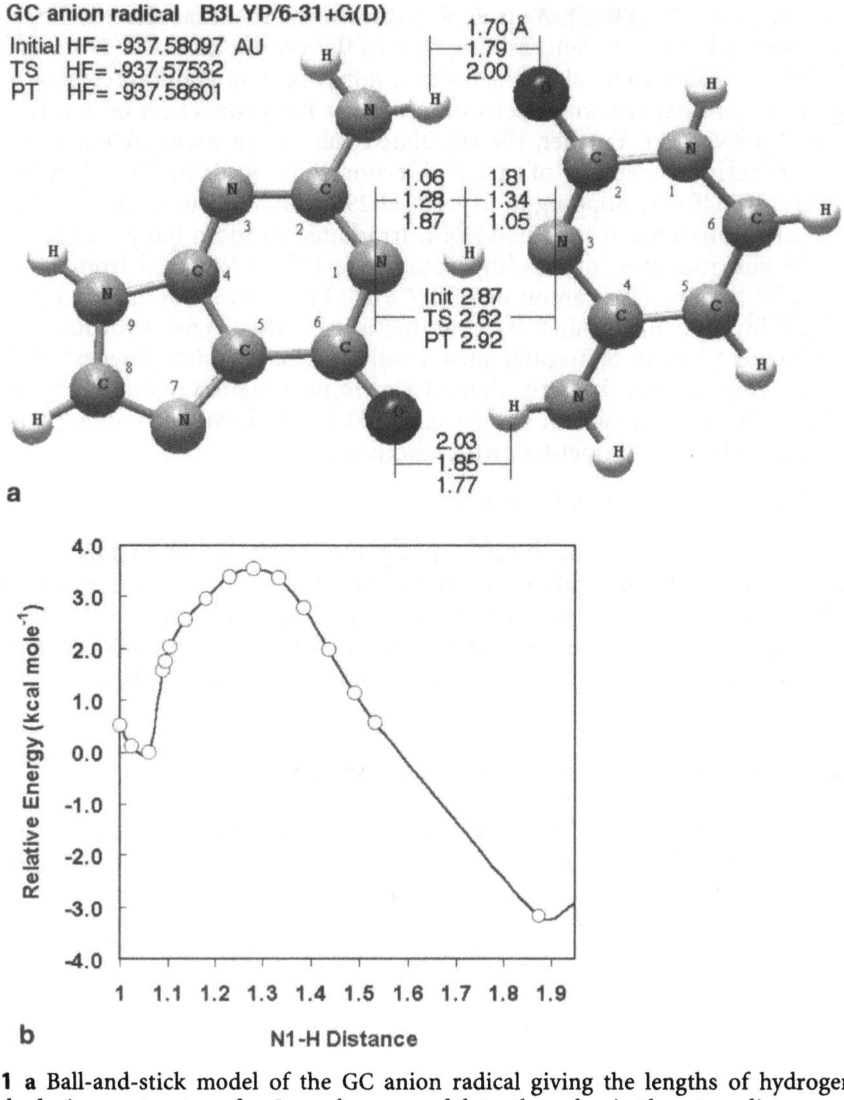

Fig. 1 **a** Ball-and-stick model of the GC anion radical giving the lengths of hydrogen bonds during proton transfer. In each group of data, the value in the *upper line* represents the initial, the *middle line* represents the transition state (TS), and the *lower line* represents the value in proton-transferred (PT) structure. **b** Density functional theory calculated (b3lyp/6-31+G(d)) energy of the GC anion radical vs increasing guanine N1-H distance [14]. Reprinted with permission from the J. Phys. Chem. Copyright (2001) American Chemical Society

Several studies [7e, 9, 10, 17] show that the first reaction (k_1) is the slow step and thus the activation energy, E_a*, for production of T(C6)H$^{\cdot}$ is the activation energy for reaction 1. Recent theoretical calculations of the barrier to reaction 1 suggest a 0.4 eV (9 kcal/mole) barrier between the base-paired protonated cytosine anion radical [C(C6)H$^{\cdot}$···G($^{-}$H)] and a base-paired thymine anion radical (T$^{\cdot -}$···A)[15]. As is expected from the fact that electron hopping (reaction 1) is the rate-determining step in the production of T(C6)H$^{\cdot}$ (reaction 2), the calculated value for electron hopping is in good agreement with the minimum experimental activation energy for production of T(C6)H$^{\cdot}$ in DNA (0.4 eV) [17]. Further, the calculated value is in reasonable agreement with the activation energy of ca. 0.3 eV reported by Anderson and Wright for electron transfer by hopping in ds DNA at 298 K in aqueous solutions [1i].

Product analyses of hydrated DNA, irradiated at room temperature, yield substantial quantities of 5,6-dihydrothymine[18] as expected from ESR results which show large amounts of T(C6)H$^{\cdot}$; however, substantial quantities of 5,6-dihydrocytosine and its deamination product 5,6-dihydrouracil are also found [Swarts S, unpublished results]. These results suggest that at higher temperatures the activation barrier to protonation of one-electron reduced cytosine at a carbon site (reaction 3) is overcome, producing a reaction path which is competitive with reaction 2.

$$\mathrm{DNA[C(N3)H^{\cdot}]} \rightarrow \mathrm{DNA[C(C6)H^{\cdot}]} \qquad (3)$$

For DNA at room temperature rapid protonation at carbon sites on both thymine and cytosine will permanently quench electron migration. Electron transfer therefore must take place in competition with these processes. At low temperatures, where irreversible protonations are prevented, the electron transfer process can be investigated and is discussed in Sect. 3.

2.4
Hole Transfer and Localization in DNA at Low Temperatures

ESR results show that hole transfer in DNA at low temperatures is initially quenched after hole localization on guanine [9] which has the lowest ionization potential of the DNA bases [15] (Scheme 2).

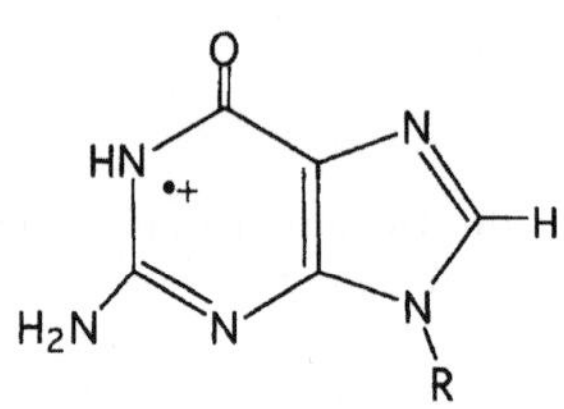

Scheme 2 Chemical structure of G$^{\cdot +}$

Hole transfer from guanine to sites of lower ionization potential such as intercalators occurs and is discussed later. ESR results also show that sugar and phosphate radicals are produced in smaller abundance [3, 9, 20–23] than expected from the 50% fraction of ionization that occurs on the sugar-phosphate backbone. This is evidence for hole transfer from the sugar ionization sites to the bases. Such hole transfer will be radioprotective in effect since sugar radicals almost invariably lead to strand breaks which, if produced in close proximity, could result in a lethal double-strand break. However, not all holes transfer to the bases. At present chemical analyses for DNA radiation damage products suggest that about half of the radiation-induced holes in the sugar-phosphate backbone transfer to the DNA bases where they predominantly end up on guanine [18, 19a]. We note that the product analysis results would imply (assuming 50% holes and 50% anions) the following relative abundance of initial radicals in DNA: 12.5% sugar radicals, 37.5% $G^{\cdot+}$,and 50% $DNA^{\cdot-}$. Our ESR results for hydrated DNA suggest 7–10% sugar radicals, 32–40% $G^{\cdot+}$, and 52–58% $DNA^{\cdot-}$ [4, 9, 20] are in generally good agreement, but suggest slightly more DNA anion radical owing to electron transfer from the hydration layer.

Product analysis shows that the oxidative pathway of direct radiation damage to DNA results in a variety of products from each of the DNA bases [18]; however, the dominant products are those from guanine such as 8-oxoguanine and fapy-guanine as expected from the overall charge transport to guanine [18].

2.5
Hole Transfer from the Hydration Layer to DNA

Transfer of radiation-induced electrons and holes ($H_2O^{\cdot+}$) from the hydration layer of DNA has been of considerable recent interest. Results from ESR experiments at low temperatures suggest that ionization of hydration water (reaction 4) results in hole transfer to the DNA (reaction 5) [4, 24–28]. Since the proton transfer reaction (reaction 6) to form the hydroxyl radical likely occurs on the timescale of a few molecular vibrations [29], it is competitive with and limits hole transfer to DNA [27].

$$H_2O \xrightarrow{\gamma} H_2O^{\cdot+} + e^{\cdot-} \qquad (4)$$

$$H_2O^{\cdot+} + DNA \rightarrow DNA^{\cdot+} + H_2O \qquad (5)$$

$$H_2O^{\cdot+} + H_2O \rightarrow H_3O^{+} + {}^{\cdot}OH \qquad (6)$$

Gregoli originally proposed that charge transfer was the primary mode of damage transfer from the hydration shell to DNA [25]. Our work [26, 27], and that of Bernhard and coworkers [28], confirms that hole transfer from water to DNA occurs but only for the first nine waters/nucleotide (Γ) which provide the contiguous surface layer. For these waters, no significant amounts of $^{\cdot}OH$ are found. For the next 12 waters/nucleotide (from Γ=9 up to Γ=21) little or no hole transfer occurs and the hydroxyl radical is found. Because these waters are associated with DNA they do not crystallize at low temperatures and the ESR spectrum appears as a hydroxyl radical in an amorphous or glassy phase, $^{\cdot}OH_{gl}$. In this amorphous hydration layer (Γ<21) no trapped electrons are found at low temperatures, which is in ac-

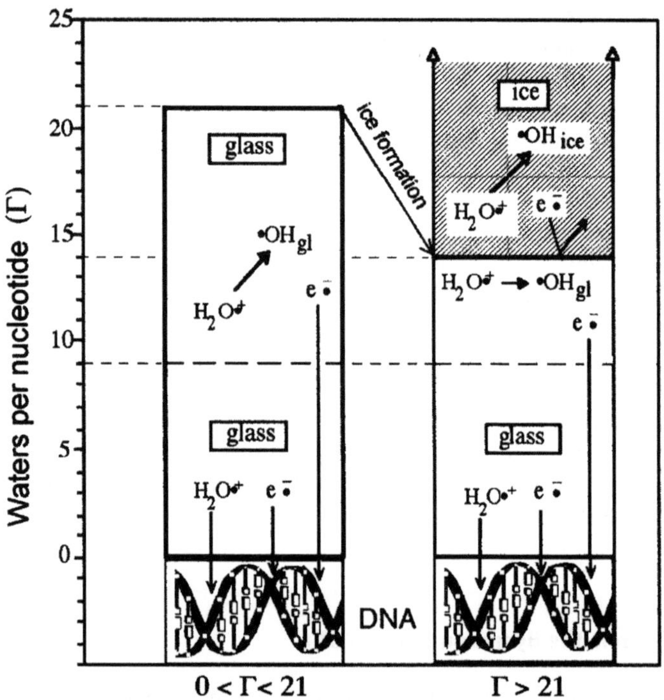

Fig. 2 Scheme depicting the radiological behavior of the waters of hydration in DNA. For the first nine waters, ESR evidence suggests that both holes and electrons efficiently transfer to the DNA. For samples with additional waters from Γ=9 to Γ=21 (Γ is defined as the number of water molecules per nucleotide), holes form hydroxyl radicals with ESR parameters characteristic of a glassy environment and electrons efficiently transfer to the DNA. For samples with Γ>21, a crystalline ice phase forms; holes in the ice phase form hydroxyl radicals with parameters characteristic of a crystalline ice environment. There is no ESR evidence that electrons from this phase transfer to the DNA. For samples with Γ>21, the glassy phase is reduced to about 14 waters per nucleotide with the remainder ice phase [3c]

cord with previous work that has reported that electrons readily transfer to the DNA from this surrounding water [4, 5, 8, 9, 30, 31] just as found in aqueous solution [32, 33] at elevated temperatures. For levels of hydration beyond 21 waters/nucleotide, a crystalline water ice phase forms on freezing which is separate from the DNA/glassy water system and neither hole nor significant electron transfer to the DNA occurs from this ice phase [4, 26, 27]. It should also be noted that when this ice phase forms(Γ>21), about seven of the glassy waters of hydration shift to the ice phase, decreasing the total number of glassy waters present to about 14 (Fig. 2) [26, 27].

Hydroxyl radicals formed in the DNA hydration shell have a high probability for reaction with the DNA and will result in DNA lesions, including base damage and strand breaks [3, 18, 19]. In this respect hole transfer from water to DNA would prevent formation of the hydroxyl radical, and may act as a radioprotective or radiosensitizing effect depending on details of the hole transfer process, i.e., whether to the sugar-phosphate backbone or to the DNA bases.

3 Electron and Hole Transfer from Trapped Ion Radical Species of DNA to Intercalators or Modified Bases

3.1 Experimental Methods

In order to investigate hole and electron transfer we have employed a number of techniques to produce holes and electron adducts within DNA [7]. These trapped ion radical species of DNA are produced by γ- or UV-irradiation at 77 K of DNA in various aqueous matrices. Each technique employed produces specific radical species.

γ-Irradiation of frozen aqueous solutions of DNA or hydrated DNA yields the guanine cation radical ($G^{\cdot+}$) and the one-electron adducts of thymine and cytosine, $CD^{\cdot}$, $T^{\cdot-}$ in the approximate amounts described in the previous section [4, 9]. Other species formed are the hydroxyl radical [26] and sugar phosphate radicals [20] but these are in smaller amounts or, in the case of hydroxyl radicals, in frozen aqueous solutions. They are mainly in a separate phase [4] and easily removed on annealing to 130 K. The few hydroxyl radicals in the glassy phase and the sugar radicals do not significantly interfere with our analyses for hole and electron transfer and are ignored.

Frozen 7M LiBr aqueous solutions containing DNA are glassy in nature and γ-irradiation at 77 K results in predominantly electrons and $Br_2^{\cdot-}$ [7a]. Electrons attach to DNA forming $CD^{\cdot}$ and $T^{\cdot-}$, whereas the holes remain trapped in the solution as $Br_2^{\cdot-}$. Photoionization of DNA is done by 254-nm UV-irradiation of frozen 8M $NaClO_4$ aqueous solutions containing DNA [34]. The process is biphotonic with ionization from a long-lived triplet state. The ionization results in $G^{\cdot+}$, and a photoejected electron which is trapped in the glass and reacts with ClO_4^- to form $O^{\cdot-}$.

Our use of intercalators applies standard techniques first employed by Peschak et al. for mitoxantrone (MX) in DNA [35]. The intrinsic binding constant and the exclusion parameter for the binding of MX to calf thymus DNA at physiological ionic strength, pH 7.0, and 25 °C were reported as $(1.85\pm0.2)\times10^5$ M^{-1} and (2.7 ± 0.2) bps [36]. MX was found to bind well to salmon testes DNA, polydAdT-polydAdT, polydIdC-polydIdC, polyA-polyU, and polyC-polyG, but poorly to polydGdC-polydGdC and polydG-polydC in 7M LiBr aqueous solution at both room temperature and 77 K [7e]. We find that intercalators appear to randomly insert within DNA if added at low ratios of intercalator to DNA base pairs [7a]. Upon irradiation of these samples, DNA electron adducts and DNA holes are trapped and over time transfer to the intercalator, which provides energetic traps for the electrons and holes. The benchmark ESR spectra of the DNA-trapped radical species and

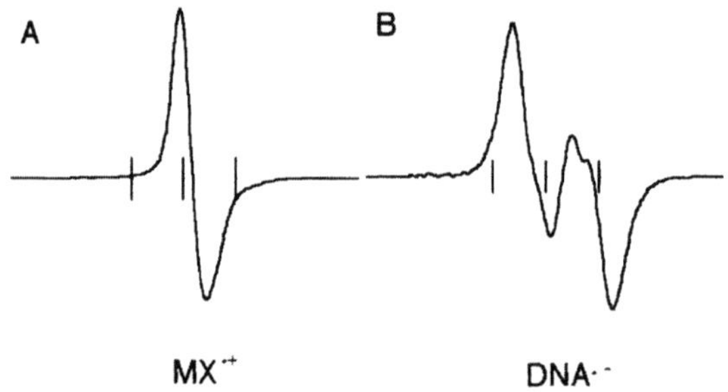

Fig. 3a,b First-derivative electron spin resonance "benchmark" spectra of **a** DNA anion radical in frozen 7M LiBr aqueous solution and of **b** one-electron reduced $MX^{(2+)}$ ($MX^{\cdot+}$) in frozen 7M LiBr aqueous solution. The three markers are each separated by 13.09 G. The central marker is at g=2.0056 [7a]. Reprinted with permission from the J. Phys. Chem. Copyright (2000) American Chemical Society

one-electron reduced or oxidized species of intercalators are obtained from model compounds. As examples in Fig. 3 we show the first-derivative ESR "benchmark" spectra of one-electron reduced DNA (a mixture of $CD^{\cdot}$ and $T^{\cdot-}$) and one-electron reduced MX radicals. These DNA and MX electron adducts were produced by γ-irradiation of frozen 7M LiBr (D_2O) glasses containing salmon sperm DNA or MX at 77 K [7a], respectively. Thus, following the ESR spectra of intercalated DNA immediately after irradiation and at time intervals of increasing length (up to weeks) allows the direct observation of electron and hole transfer in the DNA.

3.2
Analysis of Electron-Transfer Distances and Tunneling Constants

Assuming random intercalation, and that the mole ratio of intercalator or modified base to DNA base pairs (ν) is much smaller than 1, the probability that at least one of intercalator or modified base is present within distance, D, in base pairs (bp) from the site of the trapped electrons or holes is given by [37]

$$F(t) = 1 - (1 - \nu)^{2D(t)} \tag{7}$$

$F(t)$ represents the fraction of all electrons or holes captured by an intercalator or a modified base at time t relative to all electrons or holes originally captured by DNA. The diagram depicting random intercalator spacing and the scavenging range of the intercalator (D) is shown in Fig. 4. D can also be considered the electron-transfer distance to the intercalator. At high intercalator loading D has a higher probability of overlap of the scavenging ranges, as shown in the left of the diagram. The overlap at moderate loadings is found to be well accounted for by Eq. 7.

The simple rearrangement of Eq. 7 leads to the relation for the time-dependent scavenging distance along DNA double strands, $D(t)$:

$$D(t) = \frac{\ln(1 - F(t))}{2\ln(1 - \nu)} \tag{8}$$

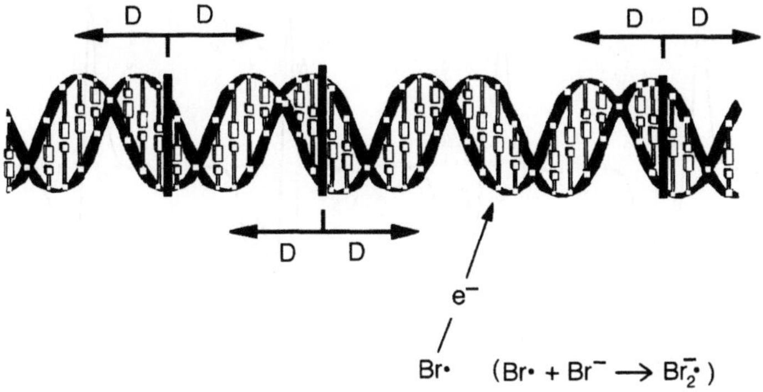

Fig. 4 Diagram depicting random intercalator spacing and the scavenging range of the intercalator (*D*). *D* is time dependent and extends outward in time. It can also be considered the electron-transfer distance to the intercalator. At high intercalator loadings the scavenging ranges have a higher probability for overlap as shown on the *left side* of the diagram. We find that this is well accounted for by Eq. 7. The electron that adds to DNA is produced by radiation of the matrix (7M LiBr, D_2O), which leaves a hole in the matrix as $Br_2\cdot^-$ [7a]. Reprinted with permission from the J. Phys. Chem. Copyright (2000) American Chemical Society

In the limit as the intercalator loading (ν) approaches zero, F also becomes small compared with 1 and the relation goes to $D=F/(2\nu)$.

Furthermore, an approximate relation for the time dependence of D, which has been successfully used for tunneling kinetics in glasses [38], is applied:

$$D(t) = (1/\beta)\ln(k_0 t) \tag{9}$$

where k_0 is the pre-exponential constant in the relation $k=k_0e^{-\beta t}$ with D and β (tunneling constant) in bp and bp^{-1}. For a tunneling process, plots of D vs $\ln(t)$ are expected to be linear with the slope equal to $1/\beta$.

Our experiments were normally carried out at 77 K. However, a detailed study of the effect of temperature was performed over the temperature range 4 to 195 K. The rates of the electron transfer in MX-DNA were found to be nearly identical from 4 to 130 K. This is discussed in more detail in Sect. 3.8.

3.3
The Effect of Electron Affinity of Intercalator

Messer et.al. studied electron transfer from the DNA anion radicals to various randomly interspaced intercalators at 77 K [7a]. The intercalators investigated included mitoxantrone (MX), ethidium bromide (EtBr), 1,10-phenanthroline (Pa), and 5-nitro-1,10-phenanthroline (NPa). As expected the fraction of electrons captured by the intercalator relative to all electrons originally trapped on DNA bases was found to increase with increased loading of intercalator. Figure 5 shows an example of first-derivative ESR spectra observed immediately after γ-irradiation of 20 mg mL^{-1} DNA in 7M LiBr

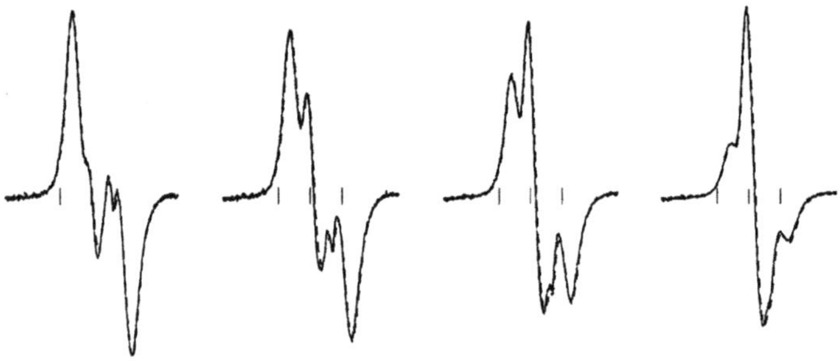

Fig. 5 First-derivative electron spin resonance spectra found immediately after γ-irradiation of samples of 20 mg/mL DNA in 7M LiBr with various loadings of MX. The *dashed* spectra are simulations made by linear least-squares fits of the benchmark functions (Fig. 3a and b) to experimental spectra. The spectra clearly show that $MX^{\cdot+}$increases in relative amount to the DNA anion radical with increased loading of MX. At the lowest loading of MX (228 bp/1 MX) these fits suggest that 8.7% of the electrons are found on MX, whereas at the highest loading (23 bp/1 MX) 59% are captured by MX with the remainder on DNA. The fraction of electrons captured by MX increases with time [7a]. Reprinted with permission from the J. Phys. Chem. Copyright (2000) American Chemical Society

with various loadings of MX. At the lowest loading of MX (228 bp/1 MX) 8.7% of the electrons are found on MX, whereas at the highest loading (23 bp/1 MX) 59% are captured by MX with the remainder on DNA.

At loadings lower than 1 MX per 20 DNA bps, the fraction of the electrons captured by the intercalator was found to follow Eq. 7 and increase with ln(t) as expected for a single-step tunneling process. The distances of electron transfer and the values of the tunneling constant for all the studied intercalators are compiled in Table 1, along with gas-phase estimates of the electron affinities (EA) of intercalators calculated by density functional theo-

Table 1 Summary of best estimates of electron- transfer distances, tunneling constants, and calculated electron affinities[a] [7a]. rReprinted with permission from the J. Phys. Chem. Copyright (2000) American Chemical Society

Intercalator	D (t=1 min) (bp)	β (Å−1)	log k_0 (k_0: s^{-1})	Calculated EA (eV)
MX	9.5±1.0	0.92±0.1	11±2	−6.25
NPa	8.3±1.0	0.85±0.2	9±3	−5.89
Pa	5.2±1.0			−5.21
EtBr	4.2±1.0	1.2±0.2	5±1	−4.32

[a] Theoretically calculated electron affinities were computed with the DFT method at the pBP86/DN*//DN level by subtraction of the total energy of the parent molecule from that of the electron adduct in the geometry of the parent molecule. The electrostatic attraction of the positively charged intercalators increases the gas- phase EA artificially. The relative values should be considered

ry. The coulombic attraction of the excess electron for the positive charges on these species accounts for the large values of the *EA*. These would be greatly diminished in a dielectric with counterions. However, the calculations give an estimate of the relative electron affinities. The electron-transfer distances of four bps (EtBr) to ten bps (MX) after 1 min at 77 K, as well as the tunneling constants β of 0.8–1.2 $Å^{-1}$, were reported. These results do not suggest that tunneling through the DNA base stack provides a particularly facile route for transfer of excess electrons through DNA—at least at low temperatures. Furthermore, the transfer distances were found to increase with increasing electron affinity of intercalators.

3.4
Electron Transfer Between DNA Duplexes

Pezeshk et al. performed the first study employing the intercalator, MX, as an electron trap in DNA [35]. They reported that the mean ET distance in frozen aqueous DNA solutions was ca. 31 bps at 77 K. In contrast, Messer et.al. reported a mean distance traveled by the electrons in frozen 7M LiBr dilute aqueous solution of DNA intercalated by MX at 77 K of only ca. 10 bps [7a]. In frozen aqueous solutions of DNA two phases are formed, one comprising pure ice and the other region containing closely packed DNA and associated water [4, 26]. In closely packed DNA at Γ=21 waters per nucleotide, the average center-to-center separation between double helices is about 24 Å [39]. In a frozen 7M LiBr aqueous solution only one glassy phase is formed, in which DNA double strands are homogeneously distributed and much better separated (the average separation between double helices is about 200 Å at 10 mg DNA/mL and 46 Å at 200 mg DNA/mL). Figure 6 is the schematic diagram depicting the spatial arrangement of DNA in a glass (fro-

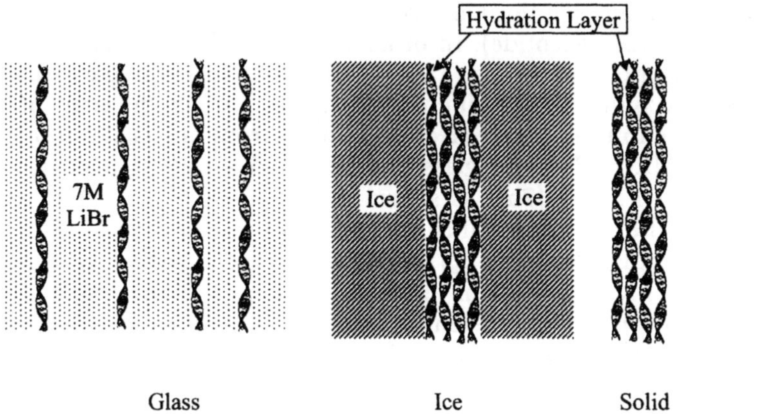

Fig. 6 Schematic diagram depicting the spatial separations of DNA duplexes in a glass (frozen 7M aqueous solution), an ice (frozen aqueous solution), and a hydrated solid (21 D_2O/nucleotide) [7b]. Reprinted with permission from the J. Phys. Chem. Copyright (2000) American Chemical Society

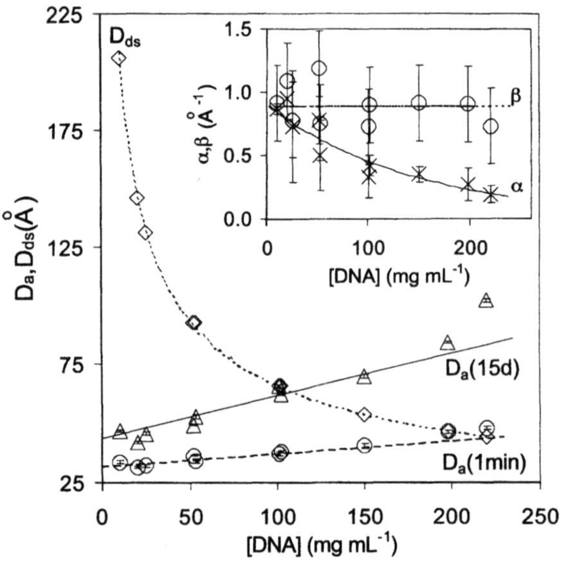

Fig. 7 Plot of apparent ET distances (D_a) at 1 min and 15 days vs concentrations of DNA in 7M LiBr glasses with loading of MX to base pair 1/52 or 1/208. The average distance between DNA double strands (D_{ds}) at each DNA concentration is also given. The lines of D_a vs [DNA] are fits to a model, which assumes transfer along a central DNA double strand and six neighboring strands. The fits are based on Eq. 10, taking D_{ds}=20 Å, D_I(1 min)=31.6 Å, and D_I(15 days)=43.9 Å. *Inserted plot* shows the dependence of apparent tunneling distance decay constants (α) on concentration of DNA in 7M LiBr glasses at 77 K. Correction for transfer to neighboring strands gives a distance decay constant, β, near one for all concentrations as shown in the *upper line in the insert* [7b]. Reprinted with permission from the J. Phys. Chem. Copyright (2000) American Chemical Society

zen 7M aqueous solution), an ice (frozen aqueous solution), and a hydrated solid (21 D_2O/nucleotide). In order to resolve the apparent discrepancy in the two works and to test a hypothesis that inter-double strand electron transfer accounts for the difference, Cai et al. studied the effect of increasing DNA concentration on the electron transfer from one-electron reduced DNA bases to MX in 7M LiBr glass and observed a clear dependence of ET on DNA concentration [7b]. As shown in Fig. 7, as the concentration of DNA increases, the average distance between DNA double strands (D_{ds}) decreases, the apparent ET distance (D_a) increases, and the apparent tunneling constant decreases. Investigations of electron transfer in frozen aqueous solutions (D_2O ices) containing different concentrations of DNA as well as in hydrated solid DNA were also performed. Significantly the apparent transfer distance and tunneling constant were found to be *independent* of the concentration of DNA when DNA water solutions were cooled to form frozen ice samples. These samples produced identical results to pure DNA which was hydrated to 21 waters/nucleotide (Γ=21). Further, the apparent transfer distances in icy and solid samples are far higher in these systems than in glassy samples.

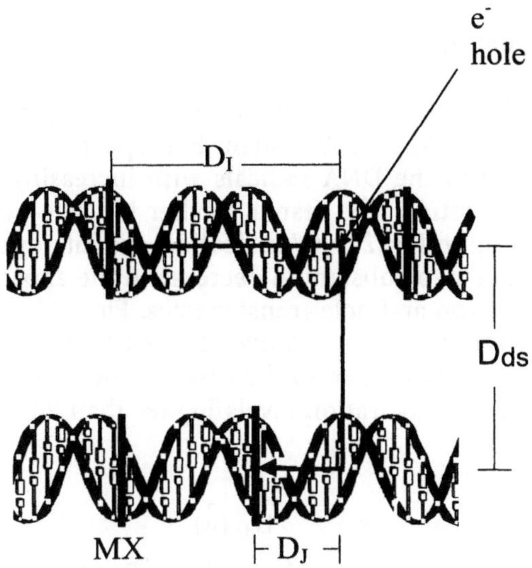

Fig. 8 Diagram depicting three-dimensional tunneling model. D_I is the total ET distance along the primary double strand. D_J is the distance of migration on the neighboring double strand; $D_J=D_I-D_{ds}$ where D_{ds} is the center-to-center interduplex distance [7b]. Reprinted with permission from the J. Phys. Chem. Copyright (2000) American Chemical Society

The values for apparent transfer distances in hydrated DNA and DNA in ices are in accord with those of Pezeshk et al. [35] and clearly indicate transfer processes occur both along the duplex and across from duplex to duplex. Bernhard and coworkers [1g] also present convincing evidence that hole/electron transfer must occur between duplexes to explain the results of their ESR work with crystalline DNA.

In order to treat solid DNA systems, Cai et al. propose a three-dimensional tunneling model that assumes electron transfer both along a primary DNA duplex and between neighboring duplexes [7b], as shown in Fig. 8. The apparent transfer distance (D_a) can be approximately related to the transfer distance along the primary DNA ds (D_I) via Eq. 10:

$$D_a(t)=D_I(t)+n(D_I(t)-D_{ds})+n(D_I(t)-1.73D_{ds}) \quad (10)$$

where $(D_I(t)-xD_{ds})$ (x=1 for first radial DNA layer or x=1.73 for the second layer) must be ≥ 0 or is set to zero; and where $D_a(t)$, $D_I(t)$, D_{ds} are in Å, n is the number of the adjacent DNA double strands at that distance and taken as 6 assuming hexagonal packing, D_{ds} is the interduplex center-to-center separation, and x is the primary duplex to surrounding duplex separation distance in D_{ds} units for each subsequent radial layer. Even though the model does not account for some of the details of the transfer, i.e., the individual strands of the double strand are distinguished but only the average of distances is employed, this model is found to fit experimental results for the concentration dependence of the apparent ET distances reasonably well (see Fig. 7).

3.5 The Effect of DNA Hydration

Cai et al. [7d] studied the effect of the level of DNA hydration on electron and hole transfer in the MX-DNA system. ESR spectra show that MX radicals decrease relative to the DNA radicals with increasing hydration levels up to Γ=22 D_2O/nucleotide. The results further indicate that, as the hydration level increases up to Γ=22 D_2O/nucleotide, the interduplex distance D_{ds} increases. This results in a substantial decrease in the apparent transfer distances as well as electron and hole transfer rates. Figure 9 shows plots of the transfer rates of electrons, holes, and overall DNA radicals at 77 K vs hydration levels (lower axis) as well as vs the distance between DNA ds's (upper axis). Please note that at hydration levels higher than 22 D_2O/nucleotide, a

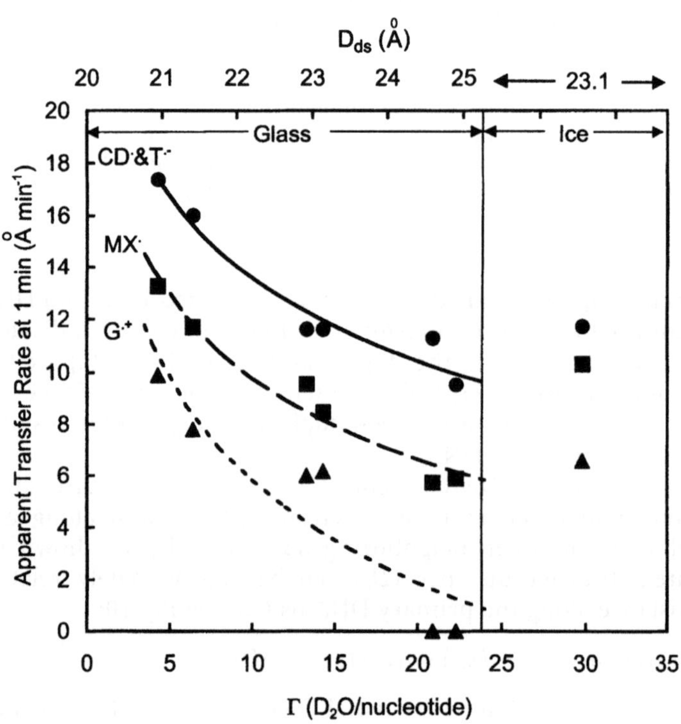

Fig. 9 Plots of the transfer rates of electrons, holes, and overall DNA radicals at 77 K vs hydration levels (*lower axis*) as well as vs the distance between DNA ds's (*upper axis*). Values of D_{ds} are estimated from the work of Lee et al. [39]. The results show that as amorphous (glassy) hydration increases up to Γ=22 D_2O/nucleotide, D_{ds} increases and transfer rate decreases. At Γ=30 D_2O/nucleotide, the ice is formed, and leaves the actual amorphous hydration level at around 14 D_2O/nucleotide with the remainder in the ice phase. The plot clearly shows equivalent transfer rates for both hydration levels at 14 and 30 D_2O/nucleotide. This result suggests that D_{ds} plays an important role in hydration-dependent hole and electron transfer in DNA [7d]. Reprinted with permission from the J. Phys. Chem. Copyright (2001) American Chemical Society

separate ice phase is formed, which steals from the amorphous phase leaving ca. 14 D_2O/nucleotide in the amorphous water layer around DNA [4, 26]. Thus near-equivalent transfer rates for the hydration levels of 14 and 30 D_2O/nucleotide are observed.

Based on the previously proposed three-dimensional model for electron tunneling both along the primary duplex and neighboring duplexes [7b], the transfer distances along the primary duplex, D_I, for the hole, $G^{\cdot+}$, excess electron, $CD^{\cdot}$, and $T^{\cdot-}$, and their average, based on the increase in MX radical, are derived from the apparent transfer distances (D_a) [7d]. This work shows that while D_a drops markedly with increasing hydration level, the transfer along the double strand, D_I, only slightly decreases with the DNA hydration level. Thus, increased separation between the DNA duplexes with hydration is proposed to be the key factor that accounts for the decreasing apparent transfer distance and rates as hydration level increases. Another less dramatic effect of the hydration layer on electron transfer is the free radical composition that at low hydration favors $T^{\cdot-}$, and at higher hydration favors the slower transferring $CD^{\cdot}$ [4a]. Also, the effect of DNA hydration level on the prototropic equilibrium within the GC cation radical base pair may affect hole transfer rates.

3.6
The Effects of Complexing Agents

Cai et al. [7d] replaced the sodium counterion of MX-DNA with various aliphatic amine cations, e.g., spermine tetrahydrochloride (SP), dodecyltrimethylammonium bromide (DOD), and octadecyltrimethylammonium bromide (OCT) and polymeric amine cations, e.g., poly-L-lysine hydrobromide (PLL) and polyethylenimine hydrochloride (PEI) to vary the separation between DNA duplexes. Molecular models of the various DNA complexes predict potential geometries and spacing (SP<PLL, PEI<DOD<OCT) induced by these complexing species in MX-DNA solid. Several are shown in Fig. 10. In hydrated complexes the apparent transfer distance and electron-transfer rates were found to decrease in the order: SP-MX-DNA>= MX-DNA>PLL-MX-DNA=PEI-MX-DNA and in dried systems they found the order to be MX-DNA>DOD-MX-DNA>OCT-MX-DNA. The radiation-produced electrons from the complexing agents readily transfer to the more electron affinic DNA. They also found that the addition of a second layer of aliphatic amine cations further suppressed the transfer of DNA radicals to near that found for isolated DNA duplex. These results support a dependence of the apparent transfer distance on the separation distance between the DNA duplexes (D_{ds}), as suggested in the 3D tunneling model. In fact, since tunneling exponentially falls off with distance, tunneling is a very sensitive measure of distances up to about 35 Å between duplexes beyond which the rates are too small to measure. Using the 3D ET model (Eq. 4) they estimated the separation distances between DNA duplexes (D_{ds}) with complexing agents from apparent distances (D_a) (Table 2). This technique was extended to chromatin in which the DNA is wrapped around the histone pro-

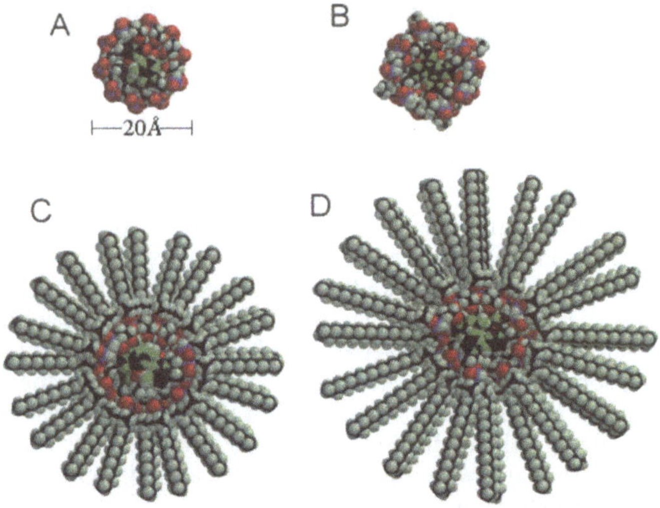

Fig. 10a–d Minimized molecular models of **a** DNA and **b** SP-DNA and idealized molecular models of **c** DOD-DNA and **d** OCT-DNA [7d]. Reprinted with permission from the J. Phys. Chem. Copyright (2001) American Chemical Society

Table 2 Results of $D_a(1')$, $D_I(1')$, and D_{ds} for overall DNA radicals transfer in various MX-DNA solid complexes [7d]

Sample	$D_a(1')$ (bps)	$D_I(1')$ (bps)	D_{ds} (Å)	State
MX-DNA	32±1	10.4±0.1[a]	23.1[c]	Hydrated
SP-MX-DNA	36±2		21±2[d]	
PEI-MX-DNA	30±2		24±1[d]	
PLL-MX-DNA	30.5±1		25±1[d]	
MX-DNA	44±2	11.1±0.4[b]	20.9[c]	Dry
DOD-MX-DNA	29±1		28±1[d]	
OCT-MX-DNA	24±1		30±1[d]	
$(DOD)_2$-MX-DNA	13±1		37±1[d]	
MX-DNA	35±2	10.8±0.3[a]	23.1[c]	Ice
MX-Nucleohistone	32±3		21±3[d]	

[a,b] $D_I(1')$ is derived from Eq. 10 using $D_a(1')$ and the D_{ds} for MX-DNA (estimated from Lee et al. [39]), respectively

[c] These values for D_{ds} are estimated from Lee et al. [39]

[d] D_{ds} is the inter-duplex separation for complexes of MX-DNA. D_{ds} is calculated from the apparent transfer distance, $D_a(1')$, which is increased by transfer between duplexes and the actual transfer distance along one duplex, $D_I(1')$, via Eq. 10, assuming $D_I(1')$ is the same for both MX-DNA and MX-DNA solid complexes at similar hydration states. All are calculated assuming $n= 6$ except MX-nucleohistone for which n was assumed to be 4.5

teins and an estimate of the distance between the strands was obtained. This work assumed that electrons formed on the histone proteins largely transferred to DNA as has been reported earlier [40, 41].

3.7 The Effect of Base Sequence

Cai et al. [7e] investigated electron and hole transfer in various polynucleotide duplexes and compared them with previous results found for salmon sperm DNA, to examine the effect of base sequence on excess electron and hole transfer along the DNA "π-way" at low temperature. Electron and hole transfer in DNA was found to be clearly base sequence dependent. In glassy aqueous systems (7M LiBr glasses at 77 K), excess electron-transfer rates increase in the order polydIdC-polydIdC<salmon testes DNA<polydAdT-polydAdT. Analogous results are found in frozen ices at 77 K where excess electron and hole transfer rates increase in the order polyC-polyG<salmon testes DNA<polyA-polyU. Transfer distances at 1 min and distance decay constants for electron and hole transfer from base radicals to MX in polynucleotides-MX and DNA-MX at 77 K are derived and compiled in Table 3. This table clearly shows that the electron-transfer rate from donor sites decreases in

Table 3 Transfer distances and distance decay constants for electron and hole transfer to MX in polynucleotides-MX and DNA-MX at 77 K [7e]

Polynucleotide	Medium	Donor site		$D_I(1')$ (bp)	Decay constants β (Å^{-1})
		ET	Hole transfer		
PolydAdT-polydAdT	D_2O glass[a]	$T^{\cdot-}$		9.4±0.5	0.75±0.1
PolydIdC-polydIdC	D_2O glass	$CD^{\cdot}$		5.9±0.5	1.4±0.1
DNA	H_2O glass	$T^{\cdot-}$+$CH^{\cdot}$		8.5±1.0	0.8±0.1
DNA[9]	D_2O glass	$T^{\cdot-}$+$CD^{\cdot}$		9.5±0.5	0.9±0.1
PolyA-polyU	D_2O ice[b]		$A^{\cdot+}$	39±10/13[c]	0.7[d]
		$U^{\cdot-}$		53±10/14[c]	0.6[d]
PolyC-polyG	D_2O ice		$G^{\cdot+}$	15±5/8[c]	1.1[d]
		$CD^{\cdot}$		35±5/11[c]	0.8[d]
DNA[10,12]	D_2O ice		$G^{\cdot+}$	17±5/8[c]	1.1[d]
		$CD^{\cdot}$+$T^{\cdot-}$		42±5/12[c]	0.7[d]

[a] Glass indicates frozen 7 M LiBr aqueous solutions

[b] Ice refers to frozen aqueous solutions which form an crystalline ice phase and regions of DNA with ca. 14 hydration waters

[c] The first value is the apparent transfer distance, $D_a(1')$, which is increased by transfer between duplexes and the second is the transfer distance along one duplex, $D_I(1')$. The latter is estimated by Eq. 10, taking n as 6 and the distance between duplexes, D_{ds}, as 23.1 Å [39]

[d] The values of β for the ice samples have been estimated from their $D_I(1')$ values assuming Eq. 9 with $k_0=1\times10^{11}$ s^{-1}. (The value of k_0 was calculated from the DNA in D_2O glass results)

the order $U^{\cdot-}$~$T^{\cdot-}$>$CH^{\cdot}$ and the hole transfer rate from donor sites decreases in the order $A^{\cdot+}$>$G^{\cdot+}$. Proton transfer from G to $C^{\cdot-}$ within the $GC^{\cdot-}$ base pair, forming $CH^{\cdot}$, and the proton transfer from $G^{\cdot+}$ to C within the $GC^{\cdot+}$ base pair resulting in contributions of $G(^{-}H)^{\cdot}C(^{+}H)^{+}$ are proposed to account in part for lower transfer distances found in polyC-polyG (with analogous processes at work in polydIdC-polydIdC). Other perhaps more important factors in electron and hole transfer processes are the primary energetics involved, that is, the adiabatic electron affinities (*EA*s) and the adiabatic ionization energies (*IE*s) of the four base pairs, as well as the relaxation energies on electron or hole transfer. Recent theoretical work using density functional theory [14, 15] has calculated these values and gives support to the experimental results found in this work. For example, the adiabatic *EA*s are predicted to increase in the order AT (0.30 eV)<AU (0.32 eV)<IC (0.42 eV)<GC (0.49 eV) and the adiabatic *IE*s to decrease in the order AT (7.68 eV)~AU (7.68 eV)>IC (7.63 eV)>GC (6.90 eV). These theoretical studies clearly show that base pairing has a profound effect on the electron localization. For example, thymine is predicted to have a 0.2 eV higher *EA* than cytosine, whereas the GC base pair has a 0.2 eV higher *EA* than AT. The sites of electron localization remain on the pyrimidines but change from thymine in individual bases to cytosine in the base pairs. In DNA, of course, other environmental influences such as waters of hydration, counterion placement, and proteins will have substantial effects as well.

3.8
The Effect of Temperature

Cai et al. [7c] employed MX-intercalated DNA and ESR to follow the time dependence of individual radical fractions as a function of temperature from 4 to 195 K in hydrated DNA and frozen glassy aqueous solutions containing DNA. By monitoring the ESR signals of MX and DNA radicals including $G^{\cdot+}$, $CD^{\cdot}$, $T^{\cdot}$, and $T(C6)D^{\cdot}$ with time, this work elucidated the ranges of temperature where tunneling, protonation, hopping or recombination are dominant (Table 4).

The rates of electron transfer in MX-DNA were found to be nearly identical from 4 to 130 K. The lack of a temperature effect suggests that tunneling

Table 4 The ranges of temperature where tunneling, protonation, hopping, or recombination are dominant in irradiated DNA samples

Dominant processes in excess electron transfer	Temperatures
Excess electron tunneling	<190 K
Activated hopping -results in recombination of $C^{\cdot-}$ and $G^{\cdot+}$	>190 K
Competing processes	
Reversible protonation of $C^{\cdot-}$ forming $C(N3)H^{\cdot}$	4 K and up
Irreversible protonation of $T^{\cdot-}$ forming $T(C6)H^{\cdot}$	>130 K
Irreversible protonation of $C^{\cdot-}$ forming $C(N6)H^{\cdot}$	>200 K

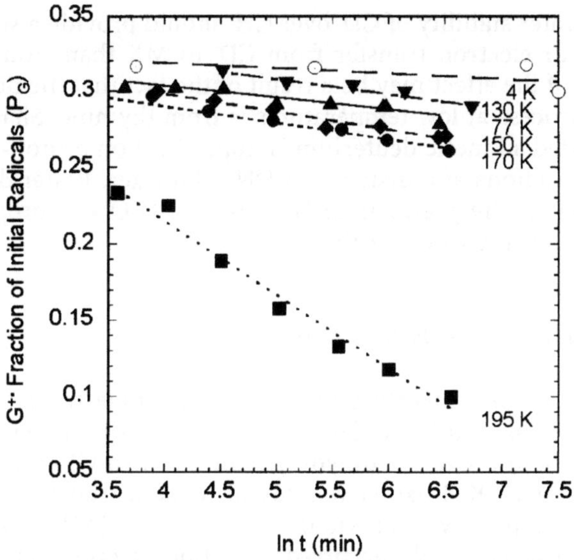

Fig. 11 Plot of the $G^{\cdot+}$ fraction (P_G) vs natural logarithm of time in min as a function of temperature. The fraction is relative to the total initial radicals formed in irradiated hydrated solid samples of MX-DNA (14 D_2O/nucleotide, ν=1/214); 4 K (*empty circles*), 77 K (*filled triangles*), 130 K (*filled upside down triangles*), 150 K (*filled diamonds*), 170 K (*filled circles*), 195 K (*filled squares*) [7c]. Reprinted with permission from the J. Phys. Chem. Copyright (2000) American Chemical Society

of electrons from DNA radicals to MX is the dominant process at low temperatures.

Work with hydrated (D_2O) DNA allows the distinction between electron adducts to cytosine and thymine which is not possible in glassy systems owing to ESR line broadening induced by the matrix. In the solid hydrated DNA below 170 K, the electron adduct to cytosine, $CD^{\cdot}$, does not undergo substantial electron loss and electron tunneling in DNA at low temperatures is suggested to be mainly from $T^{\cdot-}$. In competition with tunneling to MX, irreversible deuteration of $T^{\cdot-}$ at carbon position 6 resulting in $TD^{\cdot}$ begins at 130 K and increases in relative fraction of the total radicals as temperature increases. Hole and electron hopping resulting in recombination of $G^{\cdot+}$ and $CD^{\cdot}$ becomes substantial at temperatures near 195 K (Fig. 11). Thus, at 195 K the tunneling processes are not competitive with activated excess electron hopping.

3.9 H/D Isotope effect

Cai et al. compared electron transfer from one-electron reduced DNA base radicals to MX in D_2O glasses with H_2O glasses at 77 K [7c]. A slightly smaller value of β for electron transfer in DNA in H_2O over D_2O media was observed. The greater stability of CD˙ over CH˙ should provide a slightly weaker driving force for electron transfer from CD˙ to MX than from CH˙ to MX. The small size of the effect may be a result of the fact that the dominant electron transfer process at low temperatures is from thymine. Shafirovich et al. [42] also reported a kinetic deuterium isotope effect on proton-coupled electron-transfer reactions at a distance in DNA duplexes. Faster deprotonation and protonation of the products in H_2O than in D_2O was proposed to provide a stronger driving force for the hole transfer.

3.10 Electron and Hole Transfer to Oxidized Guanine Sites

Cai et al. [43] recently investigated electron and hole transfer to oxidized guanine bases in DNA at 77 K. Guanine in DNA was preferentially oxidized by $Br_2^{\cdot-}$ at 298 K to 8-oxo-7,8-dihydroguanine (8-oxo-G) and higher oxidation products. At 77 K 8-oxo-G was found to be a trap for radiation-produced holes in keeping with previous work of others [43]. However, further oxidation of 8-oxo-G readily occurred, resulting in products containing α-diketo groups such as oxaluric acid (G_{ox}). These oxidized species were found to be efficient electron traps. ESR spectra found after γ-irradiation of oxidized DNA samples were consistent with the presence of oxaluric acid and diketo analogues as electron traps which effectively compete with C and T for the electron. At levels of oxidation of about 1 G_{ox} to 50 bp, one-electron reduced G_{ox} [$G_{ox}^{\cdot-}$] accounts for at least half of the one-electron reduced species in γ-irradiated DNA at 77 K. As expected from duplex to duplex transfer, the fraction of $G_{ox}^{\cdot-}$ increases substantially as the center-to-center distance between nearby DNA duplexes decreases. The time profile of electron tunneling to G_{ox} was found to be a sensitive measure of level of guanine oxidation. Assuming typical values for the tunneling constant β estimates of G_{ox} levels, i.e., the G_{ox}/base pair ratio, were found which suggests a potential application of these electron-tunneling models.

3.11 Electron Transfer in Bromine-Substituted DNA: Effect of the Disruption of Base Stacking

Razskazovskiy et al. employ ESR spectroscopy at low temperatures to investigate electron transfer within brominated DNA [8]. The brominated DNA base electron traps were introduced by bromination of DNA in ice-cooled aqueous solution. The procedure is shown by NMR and GC/MS techniques to modify thymine, cytosine, and guanine, transforming them into 5-bromo-6-hydroxy-5,6-dihydrothymine, T(OH)Br, 5-bromocytosine, CBr, and 8-bromoguanine, GBr, derivatives. The bromination products formed in molar ratio close to T(OH)Br/CBr/GBr 0.2:1:0.23 and serve as internal electron scavengers on γ-irradiation. Structurally the CBr and GBr are planar, but T(OH)Br is quite nonplanar with the bromine directly above the molecular plane. This disrupts the DNA base stack. Paramagnetic products that result from electron scavenging in DNA by T(OH)Br and CBr units at 77 K were identified by ESR as the 6-hydroxy-5,6-dihydrothymin-5-yl (TOH$^{\cdot}$) radical and the 5-bromocytosine σ^* radical anion, $CBr^{\cdot-}$. Our quantitative estimates show that electron scavenging by T(OH)Br in DNA is over an order of magnitude more efficient than the more abundant CBr traps. However, in aqueous solution the reactivities are nearly equivalent. No significant trapping on GBr was found in DNA. These results indicate that within DNA, processes of electron trapping are profoundly affected by stacking. The migrating electron survives encounters with the planar CBr and GBr traps with high probability via transmission or reflection, but on encountering the disrupted DNA stack at the T(OH)Br site a reaction occurs. The displacement of the electron from its entry point is about seven bases at 77 K. The value of 11 bases reported in [8] is herein corrected for interduplex transfer, which was not recognized at the time. After trapping at 77 K migration is limited to tunneling until temperatures near 200 K, where the electron migration process (hopping) is activated and electron migration distances are found to increase substantially [8]. However, pulse radiolysis studies suggest the range of excess electron transfer to be only 3–7 bp at 300 K [1i, 33, 37a]. At these temperatures the rapid irreversible protonation processes described in Sect. 2.3 apparently severely limit the excess electron transfer range in DNA [1i, 33].

4
Concluding Remarks

Low-temperature ESR studies of irradiated DNA systems and other studies [44] have provided a clear understanding of the role of tunneling in electron transfer through DNA. Single-step tunneling is limited to transfer distances under 35 Å at a timescale of minutes. As DNA duplexes approach within this distance in solution or in the solid state, transfer between duplexes becomes competitive with transfer along the DNA duplex. Both the DNA hydration layer and positively charged molecular species can serve to separate the duplexes and retard interduplex transfer. The electron and hole tunneling through DNA is found to be clearly dependent on base sequence and the nature and energetics of the donor (DNA ion radical sites) and acceptor (trap sites). Proton transfer between base pairs is suggested to play an important role in these energetics. Thermal studies show that the depths of the electron and hole traps are overcome at temperatures near 200 K and hopping then dominates tunneling. However, at these temperatures irreversible protonations at cytosine and thymine are also activated and severely limit the range of excess electron transfer.

Acknowledgments We thank the National Cancer Institute of the National Institutes of Health (Grant RO1CA45424), and the Oakland University Research Excellence Program in Biotechnology for support of this work.

Reference

1. (a) Lewis FD, Letsinger RL, Wasielewski MR (2001) Acc Chem Res 34:159 (b) Giese B, Amaudrut J, Kohler A, Sportmann M (2001) Nature 412:318 (c) Giese B (2000) Acc Chem Res 33:631 (d) Schuster GB (2000) Acc Chem Res 33:253 (e) Sartor VBE, Schuster GB (2001) J Phys Chem A 105:11057 (f) Wan C, Fiebig T, Schiemann O, Barton JK, Zewail, AH (2000) PNAS 97:14052 (g) Debije MG, Bernhard WA (2000) J Phys Chem B 104:7845 (h) Reid GD, Whittaker DJ, Day JA, Turton DA, Kayser V, Kelly JM, Beddard GS (2002) J Am Chem Soc 124:5518 (i) Anderson RF, Wright GA (1999) Phys Chem Chem Phys 1:4827
2. (a) Berlin YA, Burin AL, Ratner MA (2001) J Am Chem Soc 123:260 (b) Berlin YA, Burin AL, Siebbeles LDA, Ratner MA (2001) J Phys Chem A 105:5666 (c) Bixon M, Jortner J (2000) J Phys Chem B 104:3906 (d) Nitzan A, Jortner J, Wilkie J, Burin AL, Ratner MA (2000) J Phys Chem B 104:5661 (e) Voityuk AA, Rösch N, Bixon M, Jortner J (2000) J Phys Chem B 104:9740 (f) Voityuk AA, Rösch N (2002) J Phys Chem B 106:3013 (g) Baik MH, Silverman JS, Yang IV, Ropp PA, Szalai VA, Yang W, Thorp HH (2001) J Phys Chem A 105:6437 (h) Smith DMA, Adamowicz L (2001) J Phys Chem B 105:9345 (i) Tong GSM, Kurnikov IV, Beratan DN (2002) J Phys Chem B 106:2381 (j) Olofsson J, Larsson S (2001) J Phys Chem B 105:10398
3. (a) Becker D, Sevilla MD (1993) The chemical consequences of radiation damage to DNA. In: Lett JT, Adler H (eds) Advances in radiation biology. Academic, New York, p 121 (b) O'Neill P, Fielden M (1993) Adv Radiat Biol 17:121 (c) Sevilla MD, Becker D (1994) Electron spin resonance (a specialist periodical report). The Royal Society of Chemistry, Cambridge 14:130 (d) Becker D, Sevilla MD (1998) Electron paramagnetic resonance (a specialist periodical report). The Royal Society of Chemistry, Cambridge, 16:79 (c) Sevilla MD, Becker D, Razskazovskii Y (1997) Nukleonika 42:283 (d) Bernhard WA (1981) Solid-state radiation chemistry of DNA: the bases. In: Lett JT, Adler H (eds) Advances in radiation biology. Academic, New York, p 199 (e) Huetterman, J (1991) Radical ions and their reactions in DNA and its constituents. Contribution of ESR spectroscopy. In: Lund A, Shiotani M (eds) Radical ions systems properties in condensed phases. Kluwer, Dordrecht, p 435 (f) Close DM (1991) Magn Reson Rev 15:241 (g) Close DM (1988) Magn Reson Rev 14:1 (h) Symons MCR (1995) Radiat Phys Chem 45:837
4. (a) Wang W, Yan M, Becker D, Sevilla MD (1994) Radiat Res 137:2 (b) Wang W, Becker D, Sevilla MD (1993) Radiat Res 135:146
5. (a) Debije MG, Milano MT, Bernhard WA (1999) Angew Chem Int Ed 38:2752 (b) Debije MG, Bernhard WA (1999) Radiat Res 152:583 (c) Milano M, Bernhard W (1999) Radiat Res 151:39
6. (a) Bernhard WA (1991) Initial sites of one-electron attachment in DNA. In: Fielden EM, O'Neill P (eds) NATO ASI Series H. Springer, Berlin Heidelberg New York, p 1415 (b) Mroczka N, Bernhard WA (1993) Radiat Res 135:155
7. (a) Messer A, Carpenter K, Forzley K, Buchanan J, Yang S, Razskazovkii Y, Cai Z, Sevilla MD (2000) J Phys Chem B 104:1128 (b) Cai Z, Sevilla MD (2000) J Phys Chem B 104:6942 (c) Cai Z, Gu Z, Sevilla MD (2000) J Phys Chem B 104:10406 (d) Cai Z, Gu Z, Sevilla MD (2001) J Phys Chem B 105:6031 (e) Cai Z, Li X, Sevilla MD (2002) J Phys Chem B 106:2755
8. Razskazovskii Y, Swarts SG, Falcone JM, Taylor C, Sevilla MD (1997) J Phys Chem B 101:1460
9. Yan M, Becker D, Summerfield S, Renke P, Sevilla MD (1992) J Phys Chem 96:1983
10. Wang W, Sevilla MD (1994) Radiat Res 138:9
11. (a) Steenken S (1997) Biol Chem 378:1293 (b) Steenken S (1989) Chem Rev 89:503
12. (a) Nelson WH, Sagstuen E, Hole EO, Close DM (1992) Radiat Res 131:10 (b) Nelson WH, Sagstuen E, Hole EO, Close DM (1998) Radiat Res 149:75 (c) Giese B, Wessely S (2001) Chem Commun 2108

13. (a) Cullis PM, Evans P, Malone ME (1996) Chem Commun 985 (b) Cullis PM, McClymont JD, Malone ME, Mather AN, Podmore ID, Sweeney MC, Symons MCR (1992) J Chem Soc Perkin Trans 2:1695
14. Li X, Cai Z, Sevilla MD (2001) J Phys Chem B 105:10115
15. Li X, Cai Z, Sevilla MD (2002) J Phys Chem B 106:9345
16. Colson AO, Besler B, Sevilla MD (1992) J Phys Chem 96:9788
17. Gräslund A, Ehrenberg A, Rupprecht A, Tjälldin B, Strom G (1975) Radiat Res 61:488
18. Swarts SG, Becker D, Sevilla MD, Wheeler KT (1996) Radiat Res 145:304
19. (a) Razskazovskiy Y, Debije MG, Bernhard WA (2000) Radiat Res 153:436 (b) Hoffmann AK, Hüttermann J (2000) Int J Radiat Biol 76:1167 (c) Swarts SG, Sevilla MD, Becker D, Tokar CJ, Wheeler KJ (1992) Radiat Res 129:333 (d) Yokoya A, Cuniffe SMT, O'Neill P (2002) J Amer Chem Soc 124:8859
20. Becker D, Razskazovskii Y, Callaghan M, Sevilla MD (1996) Radiat Res 146:361
21. Weiland B, Hüttermann J, Tol J (1997) Acta Chem Scand 51:585
22. Gatzweiler W, Hüttermann J, Rupprecht A (1994) Radiat Res 138:151
23. Close DM (1997) Radiat Res 147:663
24. Hüttermann J, Röhrig M, Köhnlein W (1992) Int J Radiat Biol 61:299
25. Gregoli S, Olast M, Bertinchamps A (1979) Radiat Res 77:417
26. La Vere T, Becker T, Sevilla MD (1996) Radiat Res 145:673
27. Becker D, La Vere T, Sevilla MD (1994) Radiat Res 140:123
28. Debije MG, Strickler MD, Bernhard WA (2000) Radiat Res 154:163
29. Klassen NV (1987) Primary products in radiation chemistry. In: Farhataziz, Rodgers MAJ (eds) Radiation chemistry. VCH, New York, p 29
30. Symons MCR (1991) The role of radiation-induced charge migration with DNA: ESR studies. In: Fielden EM, O'Neill P (eds) The early effects of radiation on DNA. Springer, Berlin Heidelberg New York, p 111
31. Cullis PM, McClymont JD, Symons MCR (1990) J Chem Soc Faraday Trans 86:591
32. Whillans DW (1975) Biochim Biophys Acta 414:193
33. Fuciarelli AF, Sisk EC, Miller JH, Zimbrick JD (1994) Int J Radiat Biol 66:505
34. Sevilla MD, D'Arcy JB, Morehouse KM, Engelhardt ML (1979) Photochem Photobiol 29:37
35. Pezeshk A, Symons MCR, McClymont JD (1996) J Phys Chem 100:18562
36. Gatto B, Zagotto G, Sissi C, Cera C, Uriarte E, Palu G, Capranico G, Palumbo M (1996) J Med Chem 39:3114
37. (a) Anderson RF, Patel KB (1991) J Chem Soc Faraday Trans 87:3739 (b) Martin RF, Anderson RF (1998) Int J Radiat Oncol Biol Phys 42:827 In this work Martin et al. present an improved representation that accounts for the introduction of the intercalator as a capture site and changes the relationship in Eq. 7 to $F(t)=1-(1-\nu)^{2D}\text{bp}^{+1}$. Thus D in Eq. 7 also contains the spacing introduced by the intercalator and can be related to the distance in bp by the relation: $D_{\text{bp}}=D-0.5$. Since this additional spacing is part of the total transfer distance we use Eq. 7 as is for our work
38. Khairutdinov RF, Zamaraev KI (1978) Russian Chem Rev 47:518
39. Lee SA, Lindsay SM, Powell JW, Weidlich T, Tao NJ, Lewen GD (1987) Biopolymer 26:1637
40. Weiland B, Hüttermann J (2000) Int J Radiat Biol 76:1075
41. Cullis PM, Jones GD, Symons MCR, Lea JS (1987) Nature 330:773
42. Shafirovich V, Dourandin A, Geacintov, NE (2001) J Phys Chem B 105:8431
43. Cai Z, Sevilla MD (2003) Radiat Res 159:411
44. Wagenknecht H-A (2003) Angew Chem Int Ed 42:2454

Top Curr Chem (2004) 237:129–157
DOI 10.1007/b94475

Proton-Coupled Electron Transfer Reactions at a Distance in DNA Duplexes

Vladimir Shafirovich · Nicholas E. Geacintov

Chemistry Department and Radiation and Solid State Laboratory, New York University, 31 Washington Place, New York, NY 10003-5180, USA
E-mail: vs5@nyu.edu
E-mail: ngl@nyu.edu

Abstract The nucleic acid analog 2-aminopurine (2AP) can be site-specifically incorporated into oligonucleotides without significantly affecting the thermal stabilities of DNA duplexes. Because the absorption band of 2AP extends beyond 300 nm, it can be selectively ionized with intense 308-nm excimer laser pulses by a tandem two-photon absorption mechanism yielding site-specifically positioned 2AP radical cations. The primary radical cations deprotonate yielding 2AP neutral radicals. These neutral radicals are strong one-electron oxidants that are capable of oxidizing guanines by a proton-coupled electron transfer mechanism even when the electron donor–acceptor pair is separated by varying numbers of bridging bases. In this process, the aqueous solution serves as a sink and source of protons in the deprotonation of the guanine radical cation and the protonation of the reduced 2AP acceptor. The involvement of solvent protons in the proton-coupled electron transfer reactions at a distance occurring within the DNA duplexes is manifested by the appearance of a solvent deuterium isotope effect on the reaction rates. The lifetimes of the guanine neutral radicals in DNA are greater than the lifetimes of their radical cation precursors and are limited by trapping reactions leading to the formation of potentially mutagenic oxidatively modified guanine bases.

Keywords DNA · Oxidative damage · Proton-coupled electron transfer · Hole transfer · Laser flash photolysis

Abbreviations and Symbols

2AP	2-Aminopurine
$2AP^{\cdot+}$	2-Aminopurine radical cation
$2AP(-H)^{\cdot}$	2-Aminopurine neutral radical
$G^{\cdot+}$	Guanine radical cation
$G(-H)^{\cdot}$	Guanine neutral radical
$dG^{\cdot+}$	2′-Deoxyguanosine radical cation
$dG(-H)^{\cdot}$	2′-Deoxyguanosine neutral radical
$dGMP(-H)^{\cdot}$	Neutral radical of 2′-deoxyguanosine 5′-monophosphate
8-oxo-dG	8-Oxo-7,8-dihydro-2′-deoxyguanosine
8-nitro-dG	8-Nitro-2′-deoxyguanosine.
e_h^-	Hydrated electron
fwhm	Full width at half of maximum

1
Oxidative DNA Damage: From Hole Injection to Oxidative Modifications of Nucleic Acid Bases and Their Biological Impact

DNA molecules in vivo are under continuous attack by free radicals, various oxidizing species, and other DNA-damaging factors. These reactions give rise to diverse, potentially mutagenic oxidative modifications (lesions) of the nucleic acid bases. The generation of DNA lesions in tissues, under conditions of oxidative stress accompanying chronic inflammation and various diseases, increases the risk of malignant cell transformation leading to the development of cancerous tumors [1, 2]. However, the harmful effects of these DNA lesions are moderated by DNA repair enzymes that excise damaged DNA bases, thus maintaining the integrity of the genetic apparatus [3]. There is growing experimental evidence that the formation of DNA lesions occurs not only at DNA bases targeted by the damaging agents, but also at nucleobases remote from these sites [4-6]. The latter effects are related to the so-called DNA chemistry at a distance [7]. These phenomena include a complex group of physicochemical processes beginning from a primary injection of electrons/holes into the DNA, migration of mobile intermediates along the DNA helix, and trapping of these intermediates at particular DNA bases, followed by their chemical transformation to stable end-products. The distribution of oxidatively modified bases along the DNA helix is a nonrandom process [4-6]. Damage of guanine, the most easily oxidizable nucleic acid base [8], occurs with greater frequency than damage to other natural DNA bases (A, C, and T). Hence, guanine can be considered as a natural trap of holes and is the most easily oxidized base in DNA.

The focus of our work and of this review is on the formation and migration of uncharged mobile intermediates (neutral radicals of nucleic acid bases) and the formation of oxidized guanine bases in DNA associated with these phenomena. Injection of holes into the DNA double helix initiates significant rearrangements of hydrogen bonding in Watson–Crick nucleic acid base pairs and surrounding water molecules, leading to the transformation of primary radical ions to neutral radicals [9]. The neutral radicals of purine bases derived from the deprotonation of the primary radical cations remain strong one-electron oxidants that can oxidize target nucleic acid bases at a distance [10–13]. In these processes, electron transfer from the target base to the neutral purine radical is coupled to a deprotonation of the donor base radical cation and the protonation of the reduced purine acceptor. The aqueous solution serves as a sink and source of protons in this deprotonation/protonation process. The involvement of solvent protons in proton-coupled electron transfer reactions is manifested by the appearance of a solvent deuterium isotope effect on the reaction rates [13, 14]. The lifetimes of typical neutral radicals in DNA are greater than the lifetimes of their precursors (radical cations/anions) and are limited by trapping reactions that ultimately lead to the formation of DNA lesions [15]. For these reasons, the oxidation processes induced by neutral radicals occur over longer distances than those associated with charge transfer reactions [11, 13]. The DNA lesions derived from these chemical transformations can give rise to mutations in living cells [1, 2].

In the next section we begin with a description of the methods developed by us for the site-selective injection of holes into double-stranded DNA [10–13].

2
Site-Selective Hole Injection in DNA Induced by Two-Photon Ionization of 2-Aminopurine

2.1
Single- and Two-Photon Excitation of 2-Aminopurine

2.1.1
Interactions of 2-Aminopurine and Nucleic Acid Bases Probed by Single-Photon Absorption

2-Aminopurine (2AP) is a structural isomer of the natural nucleic acid base adenine. This mutagenic base analog [16] forms stable Watson–Crick base pairs with thymine [17, 18] and less stable wobble base pairing with cytosine [19–21], and thus can substitute for adenine in double-stranded DNA without significantly altering the stability of the duplexes. The spectroscopic characteristics of the two aminopurine isomers, 6-aminopurine (adenine) and 2-aminopurine, are quite different. In the case of adenine in aqueous solutions, the lowest-energy absorption maximum occurs near 260 nm and the fluorescence quantum yield is only ~0.0003 [22]. In contrast, the absorption maximum of 2AP is red shifted to 305 nm, and the fluorescence emission quantum yield depends on the microenvironment and can be as high as 0.66 [23]. This particular feature of the absorption spectrum of 2AP opens an opportunity for the site-selective excitation of the 2AP residues in the presence of the normal DNA bases. The pronounced effect of the position of the amino group in the purine ring system is explained in terms of the different properties of the lowest electronic transitions in adenine and in 2AP [23]. The lowest singlet excited state of adenine has n–π^* character and decays mostly via internal conversion; on the other hand, the lowest excited state of 2AP has π–π^* character and radiative transitions of ^{1}2AP to the ground state are prominent.

The fluorescence of 2AP is strongly quenched by nucleic acid bases [17, 18, 24–29]. Time-correlated single-photon counting studies have shown that the interactions of 2AP with different nucleic acid bases significantly decrease the 2AP fluorescence lifetime [17, 24–29]. While the fluorescence lifetime of free 2AP in aqueous solution is about 10 ns, in double-stranded DNA the 2AP lifetimes are reduced to 30–50 ps. This effect has been used extensively to study the dynamics of mismatched base pairs [19, 21, 25, 30], local changes in dynamics of DNA molecules produced by their binding to the active sites of polymerases [26, 31–33], stacking interactions at abasic sites in DNA [34], and the flipping of 2AP out of the double helix in complexes with methyltransferases [35–37].

Recently, the interactions of 2AP singlet excited states with other DNA bases have been studied by pump-probe transient absorption spectroscopy with a time resolution of ~600 fs [38]. In the 2AP sequences containing either of the four natural bases (A, C, G, and T), or inosine (I):

$$5' - [2AP]A - 3' \qquad 5' - [2AP]C - 3' \qquad 5' - [2AP]G - 3'$$
$$5' - [2AP]T - 3' \qquad 5' - [2AP]I - 3' \tag{a}$$

the 2AP singlet excited states are efficiently quenched by all of these bases and the lifetimes of the 2AP singlet excited states are about 30–40 ps. The transient absorption spectra recorded in the spectral range of 320–700 nm resemble mostly the spectrum of 2AP singlet excited states. The formation of electron-transfer products is negligible if it does occur. Similar results have been obtained in the case of the 2AP-modified duplexes constructed from the single-stranded 2AP-modified oligonucleotides and corresponding complementary strands, containing T bases opposite 2AP residues

$$5' - AT[2AP]GTTTTATAAATCC - 3'$$
$$5' - AT[2AP]TTGTTATAAATCC - 3'$$
$$5' - AT[2AP]AAGTTATAAATCC - 3' \tag{b}$$

These results are not in agreement with the conclusions that guanine is the only efficient quencher of 2AP singlets in DNA duplexes [27], and that the quenching of the 2AP fluorescence by guanine occurs mostly via electron transfer from G to ^{1}2AP [28]. The single-photon excitation of 2AP in DNA does not lead to any observable formation of charge-transfer intermediates [38]. Nevertheless, excitation of the 2AP singlet excited states in DNA is useful for studies of base stacking interactions which induce a significant reduction in the 2AP fluorescence lifetime in DNA [39]. Recent fluorescence studies with femtosecond time resolution performed by fluorescence up-conversion techniques have shown that 2AP is also a sensitive probe of the hydration dynamics of DNA bases [40].

2.1.2
Consecutive Two-Photon Photoionization of 2-Aminopurine

Two-photon absorption chemistry of 2AP, specifically photoionization processes, can be induced by intense nanosecond 308-nm XeCl excimer laser pulses [10]. Typical transient absorption spectra of 2AP in deoxygenated neutral aqueous solutions are shown in Fig. 1. The stronger (385 nm) and weaker (510 nm) absorption bands were assigned to 2AP radicals derived from the ionization of 2AP (bleaching near 310 nm) [10], whereas a structureless absorption band from ~500 to 750 nm corresponds to the well-known spectrum of the hydrated electron (e_h^-) [41].

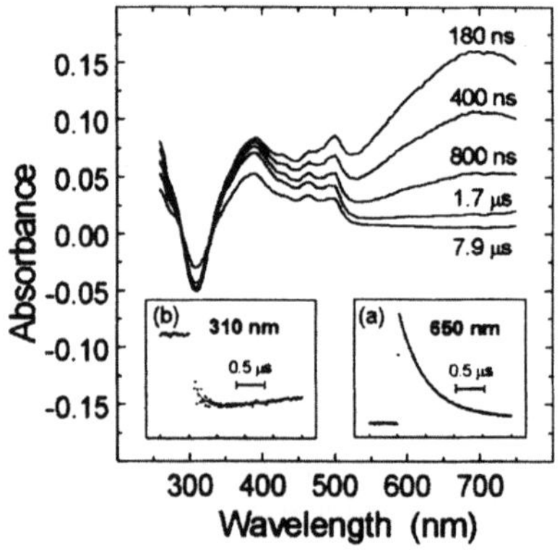

Fig. 1a,b Transient absorption spectra of 2AP (0.1 mM) in deoxygenated 20 mM phosphate buffer (pH 7) solutions recorded after 308-nm XeCl excimer laser pulse excitation (70 mJ pulse^{-1} cm^{-2}) [10]. The decay of hydrated electrons was recorded at 650 nm (**a**) and bleaching of the 2AP band at 310 nm (**b**). Reprinted with permission from the J Phys Chem, Copyright (1999) American Chemical Society

The prompt appearance of these absorption bands within a nanosecond response time of the kinetic spectrometer (Fig. 1a,b) is a direct indication that intense 308-nm laser pulses induce photoionization of 2AP [10]:

$$2AP \xrightarrow{308-\text{nm laser pulses}} 2AP^{\bullet +} + e_h^- \tag{1}$$

In deoxygenated aqueous solutions, the decay of e_h^- (Fig. 1a), besides other reaction pathways, involves the reaction of hydrated electrons with 2AP

$$2AP + e_h^- \xrightarrow{+H_2O} 2AP^{\bullet -} \rightarrow 2APH^{\bullet} + OH^- \tag{2}$$

with the rate constant (1.6×10^{10} M^{-1} s^{-1}) [10], typical of addition of e_h^- to neutral purines [42]. The radical anions of purines are unstable and are rapidly protonated by water [42].

The yields of e_h^- (Φ_e) estimated from the prompt (Δt=30 ns) signal amplitudes at 650 nm, increase nonlinearly with the laser power (I) as shown in Fig. 2. The value of n=1.75±0.1 estimated from the power dependence of Φ_e vs I^n plots is typical of a two-photon ionization process [43].

This process can be accounted for in terms of a consecutive two-photon-induced ionization. Absorption of the first photon results in the formation of the 2AP singlet excited state, and absorption of the second photon causes photoionization according to the following scheme:

$$2AP \xrightarrow{h\nu} {}^{1}2AP \xrightarrow{h\nu} 2AP^{\bullet +} + e_h^- \tag{3}$$

Previous extensive studies have shown that the energy of 193-nm photons from ArF excimer lasers, E=6.42 eV, is sufficient to induce single-photon ionization of nucleic acid bases [44–47]. The energy delivered by a consecutive two-photon excitation of 2AP is E=7.77 eV; this is the sum of the energy of the singlet excited state of ^{1}2AP (E_{00}=3.74 eV) and the energy of a 308-nm

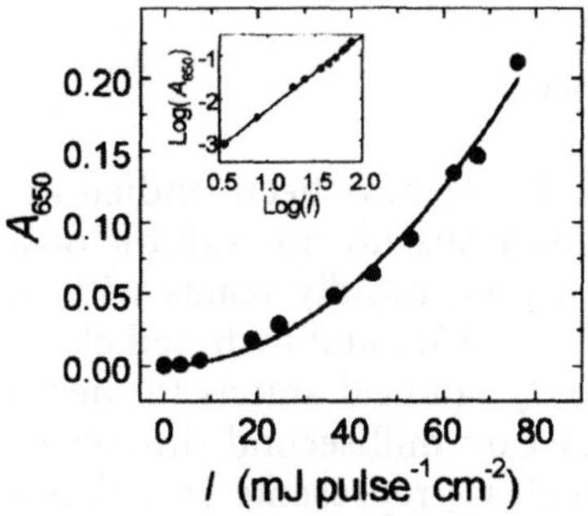

Fig. 2 Laser power curves for yields of hydrated electrons (transient absorption signal measured at 650 nm with a delay time of 30 ns) generated with 308-nm XeCl excimer laser pulse excitation of 2AP in deoxygenated 20 mM phosphate buffer (pH 7) solutions [10]. Reprinted with permission from the J Phys Chem, Copyright (1999) American Chemical Society

photon (E=4.03 eV). This total energy is greater than the energy of a single 193-nm photon, and it is therefore not surprising that irradiation of 2AP with intense 308-nm laser pulses causes photoionization.

2.2 Reaction Pathways of 2-Aminopurine Radicals

2.2.1 *Acid–Base Equilibria in Aqueous Solutions*

The purine radical cations are strong Brønsted acids, and thus rapidly deprotonate in neutral aqueous solutions [47, 48]. For instance, the adenosine radicals generated by reaction with $SO_4^{\cdot-}$ radicals were identified as neutral radicals, A(-H)$^{\cdot}$. A change in pH from 1 to 13 does not affect the transient absorption spectra of these species; based on this result, it was inferred that the pK_a of $A^{\cdot+}$ is $\leq$1 [49]. The solution pH exerts pronounced effects on the transient absorption spectra of the 2AP radicals assigned to the following equilibrium:

$$2AP^{\bullet+} \rightleftharpoons 2AP(-H)^{\bullet} + H^{+} \quad (4)$$

with pK_a=2.8±0.2 [50]. Using this pK_a value and assuming that protonation of 2AP(-H)$^{\cdot}$ occurs with a diffusion-controlled rate constant (~2× 10^{10} M^{-1} s^{-1}) [48], we estimate that the deprotonation rate constant of $2AP^{\cdot+}$ is ~3×10^7 s^{-1} [10], i.e., the lifetime of $2AP^{\cdot+}$ is about 30 ns. Therefore, we conclude that the transient absorption spectra in the ~330–500 nm range observed on microsecond timescales (Fig. 2) are mostly due to 2AP(-H)$^{\cdot}$ neutral radicals rather than to $2AP^{\cdot+}$ radical cations.

2.2.2
Oxidation of Nucleic Acid Bases

Oxidative reactions of 2AP(-H)$^{\cdot}$ have been studied in oxygen-saturated solutions because 2AP(-H)$^{\cdot}$ radicals do not exhibit observable reactivities towards O_2 [10]. In contrast, O_2 rapidly reacts with hydrated electrons (rate constant of 1.9×10^{10} M^{-1} s^{-1}) [51] and hydrated electrons do not interfere in oxidative processes. The $O_2^{\cdot-}$ radical anions formed do not react with intact nucleic acid bases, at least on millisecond timescales [47]. Thus, saturating the aqueous solutions with O_2 represents an efficient method for reducing the efficiencies of side reactions, such as Eq. 2.

The 2AP(-H)$^{\cdot}$ radicals selectively oxidize guanine [10] and 8-oxoguanine [12], the latter representing an important marker of DNA oxidation in vivo [1, 52]. Elevated levels of 8-oxo-dG have been detected in vivo under conditions of oxidative stress accompanying inflammation and characterizing various diseases [53–55]. In contrast, 2AP(-H)$^{\cdot}$ radicals do not react with other 2′-deoxynucleosides (dA, dC, dT) or nucleotides to any observable extent [10, 12]. This trend in the reactivities of 2AP(-H)$^{\cdot}$ is associated with different redox potentials of nucleic acid bases. The redox potentials of dG, E_7[dG(-H)$^{\cdot}$/dG]=1.29 V vs NHE [8], and 8-oxo-dG, E_7[8-oxo-dG(-H)$^{\cdot}$/8-oxo-dG]=0.74 V vs NHE [56], are lower than the redox potentials of dA, E_7[dA(-H)$^{\cdot}$/dA]=1.42 V vs NHE, and of the pyrimidine bases, E_7=1.7 V (dT) and 1.6 V (dC) vs NHE [8].

2.2.3
Solvent Kinetic Deuterium Isotope Effects

The remarkable solvent isotope effect on the kinetics of oxidation of guanine by 2AP radicals has been detected in H_2O and D_2O solutions [14]. In H_2O, the rate constants of G(-H)$^{\cdot}$ formation are larger than those in D_2O by a factor of 1.5–2.0 (Table 1). This kinetic isotope effect indicates that the electron transfer reaction from guanine to 2AP radicals is coupled to deprotonation/protonation reactions of the primary electron-transfer products (Scheme 1).

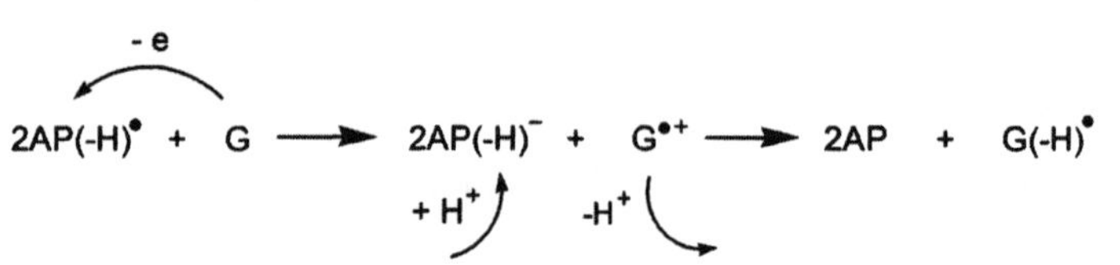

Scheme 1 Proton-coupled electron transfer from guanine to 2AP neutral radical

These proton transfer processes increase the driving force of the electron transfer reactions, which can thus be considered in terms of a proton-coupled electron transfer process [57–60].

Table 1 Deuterium isotope effect on the rate constants of oxidation of 2′-deoxyguanosine 5′-monophosphate by one-electron oxidants in neutral aqueous solutions (adapted from [14] and [61])

One-electron oxidant	k (10^7 M^{-1} s^{-1})		
	H_2O	D_2O	H_2O/D_2O
2APr(-H)[a]	6.2±0.6	2.9±0.3	2.1±0.3
2AP(-H)˙	3.5±0.4	2.3±0.2	1.5±0.2
$BPT^{\cdot+}$	170±10	110±10	1.6±0.2
($BPT^{\cdot+}$...dAMP)	32±2	22±2	1.5±0.2

[a] 2APr is 2-amino-9-*β*-D-ribofuranosylpurine

The solvent kinetic deuterium isotope effect on the rates of guanine oxidation is typical of its reactions with other one-electron oxidants. Thus, the reactions of the radical cation of a pyrene derivative, 7,8,9,10-tetrahydroxytetrahydrobenzo[*a*]pyrene (BPT), or of the noncovalent complex of $BPT^{\cdot+}$ with dAMP (Table 1), exhibit deuterium isotope effects [61]. The kinetic isotope effects observed for photoinduced electron transfer reactions in covalently linked benzo[*a*]pyrene diol epoxide–guanosine adducts [62], noncovalent benzo[*a*]pyrenetetrol–nucleoside complexes [63], electrocatalytic oxidation of guanine in DNA [64], and excess electron migration in DNA [65] provide strong support for the notion that these reactions are all coupled with a proton transfer process. Experimental and theoretical studies showed that proton-coupled electron transfer reactions can occur in DNA–acrylamide complexes [66, 67], and in the photooxidation occurring in noncovalent guanine–cytosine complexes by *N*,*N*′-dibutylnaphthaldiimide or fullerene in the triplet excited states [68, 69].

2.3 Two-Photon Ionization of 2-Aminopurine in Single- and Double-Stranded DNA

In aqueous solutions, the photoexcitation of the 2AP-modified oligonucleotides with intense nanosecond 308-nm excimer laser pulses results in the site-selective two-photon ionization of the 2AP residues [10, 11, 13]. Absorption of the first photon results in the formation of the 2AP singlet excited state, and the absorption of a second photon causes the photoionization of ^{1}2AP as shown in Eq. 3. The lifetime of free ^{1}2AP in aqueous solution is 10 ns [24] and favors absorption of the second photon delivered by the same excimer laser pulse (fwhm=12 ns, 10 pulse s^{-1}). When 2AP is base-paired with any of the four normal bases in duplex DNA, the mean lifetime of ^{1}2AP is shortened significantly [17, 18, 24–29], thus lowering the yield of two-photon photoionization [10]. Indeed, there is a correlation between the e_h^- yield ratios, $\Phi_e(\text{oligo})\Phi_e(\text{free})^{-1}$, and the fluorescence yield ratios, $F(\text{oligo})F(\text{free})^{-1}$, which are proportional to the mean lifetimes of ^{1}2AP in the oligonucleotides and free solutions [10]. For these reasons, we positioned the 2AP residues at the 5′-ends of the oligonucleotides [11–13], because fraying of the duplexes at the ends allows for some dynamic motion, thus increasing the lifetimes of ^{1}2AP. Indeed, the e_h^- and fluorescence yield ratio are enhanced when 2AP is positioned at the ends of the duplexes rather than in the middle [10, 11]. In contrast, the photoexcitation of the oligonucleotides containing no 2AP bases with intense nanosecond 308-nm laser pulses (~70 mJ pulse^{-1} cm^{-2}) does not induce photoionization of other normal nucleic acid bases (A, C, G, and T) [10, 11, 13]. The singlet excited states of the normal DNA bases (A, C, G, and T) decay on ps timescales with negligible yields of the triplet excited states [70], so the probability of the absorption of a second photon during a nanosecond laser pulse becomes very low [71, 72]. These features provide the opportunity for photoionizing 2AP without ionizing any of the normal DNA bases, and the site-selective generation of 2AP radical cations or radicals in double-stranded oligonucleotides.

3 Oxidation of Guanine at a Distance in DNA Induced by 2-Aminopurine Radicals

3.1 Design of 2-Aminopurine-Modified Duplexes

The effects of distance on the oxidation of guanine by 2AP radicals were studied using complementary duplexes (with T opposite 2AP) with the following 15-mer oligonucleotides, each containing a single 2AP base at the 5′-end [11, 13].

$$
\begin{aligned}
&5' - [2AP]GGTTTTTTTTTTTT - 3'\\
&5' - [2AP]GGAAAAAAAAAAAA - 3'\\
&5' - [2AP]TGGTTTTTTTTTTT - 3'\\
&5' - [2AP]AAGGAAAAAAAAAA - 3'\\
&5' - [2AP]TTGGTTTTTTTTTT - 3'\\
&5' - [2AP]AAAAGGAAAAAAAA - 3'\\
&5' - [2AP]TTTGGTTTTTTTTT - 3'\\
&5' - [2AP]AAAAAAGGAAAAAA - 3'\\
&5' - [2AP]TTTTGGTTTTTTTT - 3'\\
&5' - [2AP]AAAAAAAAGGAAAA - 3'\\
&5' - [2AP]TTTTTTTTTTTTTT - 3'\\
&5' - [2AP]AAAAAAAAAAAAAA - 3'
\end{aligned}
\qquad (c)
$$

The $[2AP]A_nGGA_{12-n}$ and $[2AP]T_nGGT_{12-n}$ oligonucleotides containing a single 2AP residue and a single GG doublet separated only by adenine (n=0, 2, 4, 6, 8) or thymine (n=0, 1, 2, 3, 4) bases were used to study base sequence effects on electron transfer at a distance. The $[2AP]A_{14}$ and $[2AP]T_{14}$ oligonucleotides were used as controls. All of the duplexes exhibit well-defined cooperative melting behavior with melting temperatures in the range T_m=39–44 °C. The melting curves measured within the absorption band of DNA (near 260 nm) resemble those recorded within the 2AP absorption band at 310 nm, and the T_m values are the same, within experimental error. Thus, even though the 2AP residues are positioned at the end of the duplexes, the dynamic fraying at the ends does not significantly influence the cooperativity of duplex dissociation.

3.2
Heterogeneous Kinetics of Guanine Oxidation

Based on our results discussed below, we assume that in DNA duplexes, the decay of $2AP^{\cdot+}$ radical cations generated by the 308-nm laser pulses occurs mostly via two competitive pathways. These two pathways include deprotonation of $2AP^{\cdot+}$ with the formation of the neutral radical, $2AP(\text{-H})^{\cdot}$, and hole transfer from $2AP^{\cdot+}$ to nearby guanines. The characteristic time of the $2AP^{\cdot+}$ deprotonation in DNA duplexes is expected to be close to that of free $2AP^{\cdot+}$ (~30 ns) [10]. The N2 proton in $2AP^{\cdot+}$ does not participate in Watson–Crick hydrogen bonding with the complementary T base and hence, the $2AP^{\cdot+}$ deprotonation does not require a base pair opening event as in the case of the exchange of imino protons of Watson–Crick base pairs in DNA [73–75]. The lifetimes of $2AP^{\cdot+}$ radical cations in DNA duplexes may be even shorter than 30 ns, since $2AP^{\cdot+}$ can also decay via a competitive hole transfer reaction to neighboring guanine bases.

The neutral $2AP(\text{-H})^{\cdot}$ radical is also a strong one-electron oxidant of guanine bases in DNA [10, 11, 13]. The time window for the observation of guanine oxidation by $2AP(\text{-H})^{\cdot}$ at a distance is expected to be in the range of ~30 ns (deprotonation of $2AP^{\cdot+}$) to ~0.5 ms (decay of $2AP(\text{-H})^{\cdot}$ in side reactions). The existence of two potential one-electron oxidants ($2AP^{\cdot+}$ and $2AP(\text{-H})^{\cdot}$) with very different lifetimes suggests that heterogeneous decay kinetics of guanine radical formation in DNA might be observable on nanosecond timescales. Our experiments reveal several different time-dependent and sequence-dependent phases of guanine radical formation in (~100 ns) nanosecond to microsecond time regimes, suggesting that other factors may govern the observation of nonexponential electron transfer kinetics as well [11, 13]. In the next section we begin with discussions of the electron transfer kinetics in oligonucleotides with thymidine bridging bases between the 2AP and guanines. In the following section we discuss the case of adenine bridging bases.

3.2.1
Formation of Guanine Radicals in Duplexes with T Bridging Bases

Representative transient absorption spectra of duplexes with [2AP] T_nGGT_{12-n} in O_2-saturated neutral solutions are summarized in Fig. 3. The decay of $2AP^{\cdot+}$ radical cations via deprotonation and the oxidation of gua-

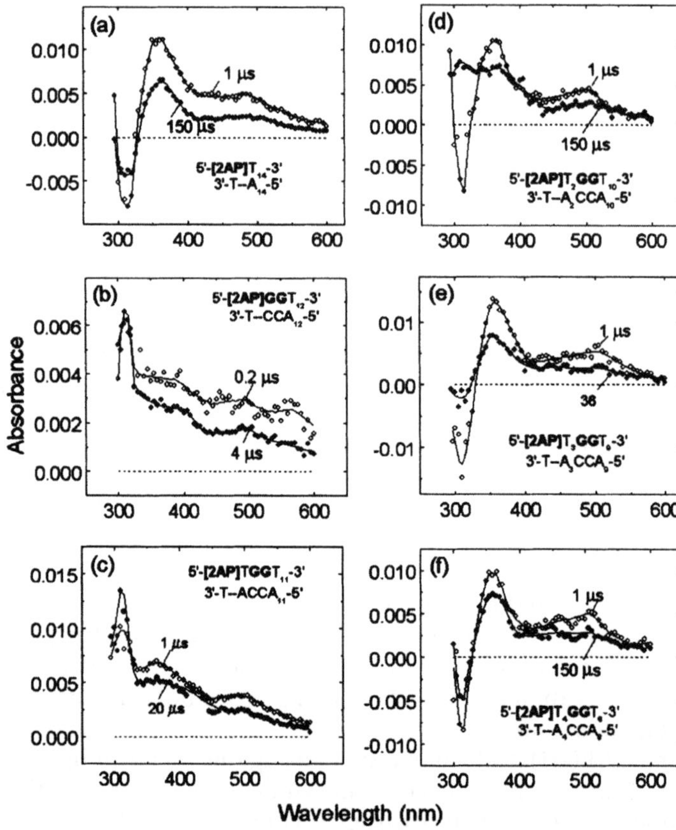

Fig. 3 Transient absorption spectra of the $[2AP]T_nGGT_{12-n}$ duplexes (0.1 mM) in oxygenated 20 mM phosphate buffer (pH 7) solutions recorded after 308-nm XeCl excimer laser pulse excitation (60 mJ pulse^{-1} cm^{-2}) [11]. Reproduced by permission of The Royal Society of Chemistry on behalf of the PCCP Owner Societies

nine were not time-resolved in these experiments because of scattered laser light and temporary photodetector overload which limits our observation times to greater than ~100 ns. However, the primary products of the $2AP^{\cdot+}$ decay, the $2AP(\text{-H})^{\cdot}$ and $G^{\cdot+}/G(\text{-H})^{\cdot}$ radicals, are clearly observable in the transient absorption spectra measured at ~100 ns and longer time domains.

The spectra of the $[2AP]T_{14}$ duplexes containing no guanines (Fig. 3a) are characterized by a bleaching of the 2AP absorption band near 310 nm, and the appearance of the absorption bands of the $2AP(\text{-H})^{\cdot}$ neutral radicals; in DNA, the absorption maxima of these bands are observed at 365 nm (strong) and 510 nm (weak) [11]. The spectra obtained with the $[2AP]GGT_{12}$ duplexes containing no intervening bases between the electron donor and acceptor residues (Fig. 3b) exhibit the characteristic narrow absorption band near 310 nm typical of the $G^{\cdot+}$ or $G(\text{-H})^{\cdot}$ species with similar absorption spectra [48]. In our work, the formation of the $G^{\cdot+}/G(\text{-H})^{\cdot}$ species occurs on timescales faster than the time resolution of our experimental setup ($\leq$100 ns), and the difference between the spectra recorded at Δt=200 ns and 4 μs (Fig. 3b) is attributed to the tail of the hydrated electron absorption band [41]. In the [2AP]T_1GGT_{11} duplex (Fig. 3c), the major component (~85% of the total absorbance) of the $G^{\cdot+}$/G(-H)$^{\cdot}$ transient absorbance was not resolved in time ($\leq$100 ns); however the decay time of the minor (~15%) component was determined to be ~2 μs. In the [2AP]T_2GGT_{10} duplex, the contribution of the fast component ($\leq$100 ns) decreases to ~25% and the slower phase of formation of the guanine radicals on a timescale of 0.1 ms is clearly evident (Fig. 3d). The fast component was not observable in the other duplexes [2AP]T_nGGT_{12-n} with n=3 and 4, and the formation of guanine radicals occurs entirely in the 0.2ms time intervals (Fig. 3e,f).

Analysis of the transient absorption spectra recorded at 100 ns allows for the determination of the prompt relative yields of the $G^{\cdot+}$/G(-H)$^{\cdot}$ radicals, Φ_G ($\leq$100 ns). This yield is probably due, in part, to the oxidation of the guanines by the $2AP^{\cdot+}$ radical cations [11]. Assuming that the $2AP^{\cdot+}$ radical cations decay only via the deprotonation of $2AP^{\cdot+}$ (rate constant k_h) and hole transfer from $2AP^{\cdot+}$ to guanine (k_t), the prompt yield Φ_G may be expressed as follows:

$$\Phi_G = k_h/(k_h + k_t) = [G^{\bullet+}/G(-H)^{\bullet}]_{t=100}/\big([G^{\bullet+}/G(-H)^{\bullet}]_{t=100} + [2AP(-H)^{\bullet}]_{t=100}\big) \quad (5)$$

where $[G^{\cdot+}/G(-H)^{\cdot}]_{t=100}$ and $[2AP(-H)^{\cdot}]_{t=100}$ are the concentrations of the radicals measured at a delay time of 100 ns; at this delay time, all of the $2AP^{\cdot+}$ radical cations have been converted to the neutral radical 2AP(-H)$^{\cdot}$. The values of Φ_G as a function of the number of intervening thymidine residues are summarized in Fig. 4. In the [2AP]GGT_{12} duplexes containing no intervening bases between the electron donor and acceptor residues, the values of Φ_G are close to 1 and there is no observable slow phase of G(-H)$^{\cdot}$ formation. The values of Φ_G sharply diminish as the number of the intervening bases

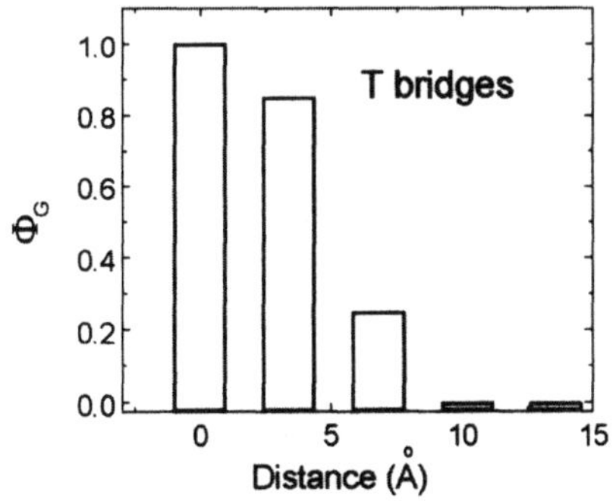

Fig. 4 The prompt (Δt=100 ns) yields of the $G^{\cdot+}$/G(-H)$^{\cdot}$ radicals, Φ_G, as a function of the number of T bridging bases in the double-stranded oligonucleotides (adapted from [11])

between the 2AP and GG residues is increased, and Φ_G approaches zero in the duplex with the bridging TTT sequence.

3.2.2
Formation of Guanine Radicals in Duplexes with A Bridging Bases

The kinetics of $G^{\cdot+}/G(-H)^{\cdot}$ formation in the $[2AP]A_nGGA_{12-n}$ duplexes are even more heterogeneous than in the case of bridging thymidines [11, 13]. For instance, in the $[2AP]GGA_{12}$ duplexes, as in the $[2AP]GGT_{12}$ duplexes, the formation of the $G^{\cdot+}/G(-H)^{\cdot}$ radicals occurs on timescales faster than the time resolution of our experimental setup ($\leq$100 ns). The slow millisecond components were detected in the $[2AP]A_nGGA_{12-n}$ (n=6 and 8) as well as in the $[2AP]T_nGGT_{12-n}$ (n=3, 4) duplexes. In the $[2AP]A_nGGA_{12-n}$ (n=2, 4) duplexes, the kinetics of $G^{\cdot+}/G(-H)^{\cdot}$ formation contains at least three components with wide differences in time constants.

The transient absorption spectra of duplexes with $[2AP]A_4GGA_8$ are depicted in Fig. 5. At a delay time of 100 ns, the transient absorption spectrum is attributed to the superposition of the spectra of the $2AP(-H)^{\cdot}$ and $G^{\cdot+}/G(-H)^{\cdot}$ radical products and the hydrated electrons. The structureless tail of the e_h^- absorption in the 350–600 nm region decays completely within Δt<500 ns. The formation of $G^{\cdot+}/G(-H)^{\cdot}$ radicals monitored by the rise of the 310-nm absorption band and associated with the decay of the $2AP^{\cdot+}$/ $2AP(-H)^{\cdot}$ transient absorption bands at 365 and 510 nm (Fig. 5) occurs in at least three well-separated time domains (Fig. 6). The prompt ($\leq$100 ns) rise of the transient absorption at 312 nm due to guanine oxidation by $2AP^{\cdot+}$ was not resolved in our experiments. However, the amplitude, $A_{(t=100)}$, related to the prompt formation of the $G^{\cdot+}/G(-H)^{\cdot}$ radicals (Fig. 6a) can be estimated using the extinction coefficients of the radical species at 312 and 330 nm (isosbestic point) [11]. The kinetics of the $G^{\cdot+}/G(-H)^{\cdot}$ formation in the μs and ms time intervals were time-resolved and characterized by two well-defined components shown in Fig. 6a (0.5 μs) and Fig. 6b (60 μs).

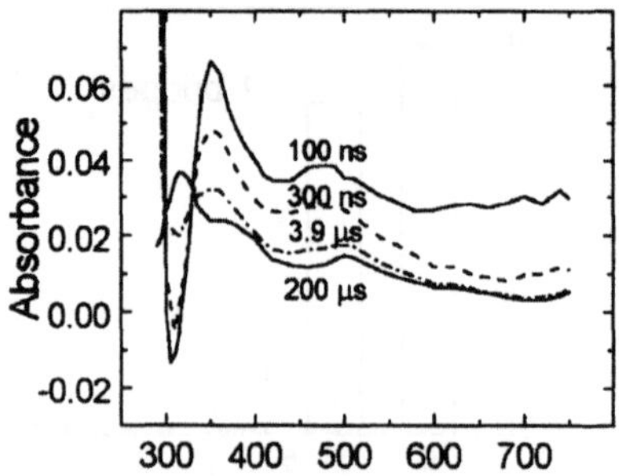

Fig. 5 Transient absorption spectra of the $[2AP]A_4GGA_8$ duplex (0.1 mM) in oxygenated 20 mM phosphate buffer (pH 7) solutions recorded after 308-nm XeCl excimer laser pulse excitation (60 mJ pulse^{-1} cm^{-2})

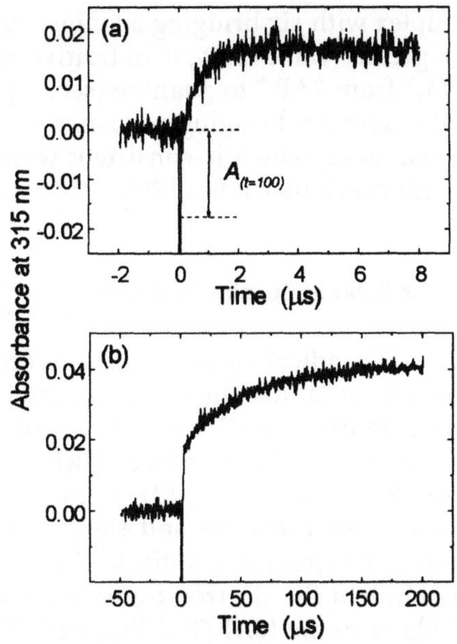

Fig. 6 Kinetics of the $G^{\cdot+}/G(-H)^{\cdot}$ formation in the [2AP]A_4GGA_8 duplex (adapted from [11]). The *solid lines* are best single-exponential fits to the experimental data points. Other conditions are shown in Fig. 5

Thus, the kinetics of $G^{\cdot+}/G(-H)^{\cdot}$ formation in the [2AP]A_nGGA_{12-n} duplexes reveal a more heterogeneous character than in the [2AP]T_nGGT_{12-n} duplexes.

The relative contribution of the fast processes of $G^{\cdot+}/G(-H)^{\cdot}$ formation in [2AP]A_nGGA_{12-n} duplexes is shown in Fig. 7. In the [2AP]A_nGGA_{12-n} duplexes, the decrease of Φ_G as a function of the number of bridging bases is slower than in the case of the [2AP]T_nGGT_{12-n} duplexes (Fig. 4), and Φ_G ap-

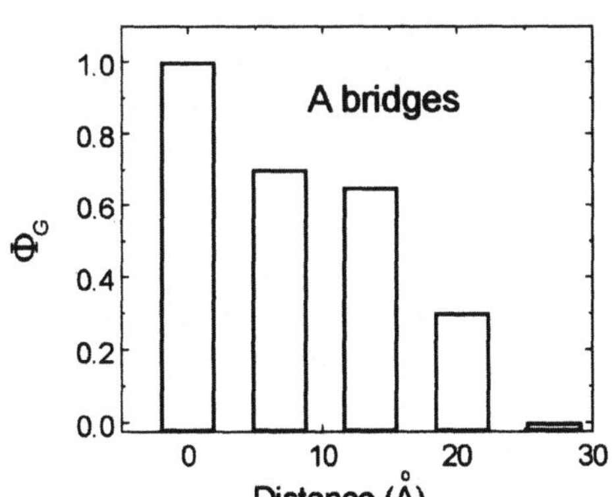

Fig. 7 The fast ($\Delta t \leq 2\ \mu s$) yields of the $G^{\cdot+}/G(-H)^{\cdot}$ radicals, Φ_G, as a function of the number of A bridging bases in double-stranded oligonucleotides

proaches zero in the duplex with six bridging adenine bases (n=6). At a fixed number of bases (n), a greater value of Φ_G is indicative of a greater rate constant of hole transfer (k_t) from $2AP^{\cdot+}$ to guanine (see, Eq. 5). The hole transfer reaction through the adenine bridging sequence is faster than through the thymine bridging sequence, which is consistent with theoretical considerations [76, 77] and with previous studies [78].

3.3 Base Sequence Effects on the Rates of Guanine Oxidation

The deprotonation of $2AP^{\cdot+}$ radical cations does not seem to interrupt the DNA oxidation chemistry at a distance. The neutral radical, 2AP(-H)$^{\cdot}$, formed via a proton release from $2AP^{\cdot+}$, remains a strong one-electron oxidant that can oxidize guanine at the level of free nucleosides [10, 14, 50] and single- and double-stranded DNA [10, 11, 13]. In the case of the [2AP]A_6G-GA_6 and [2AP]A_8GGA_4 duplexes with six and eight bridging adenine bases, the transient absorption measurements indicate that guanine oxidation by 2AP(-H)$^{\cdot}$ is dominant, and that the electron transfer rate constants are of the same order of magnitude as in the [2AP]T_2GGT_{10} and [2AP]T_3GGT_9 duplexes that have only two and three bridging thymidines. In all cases, the first-order electron transfer processes from GG to 2AP(-H)$^{\cdot}$ are intraduplex processes since these rate constants of electron transfer are independent of the concentrations of duplexes in the 10–100 μM range [13]. These systems are therefore excellent objects for exploring intraduplex electron transfer reactions at a distance in double-stranded DNA.

The general time dependence of the transient absorption signals, $A(t)$, recorded at two representative wavelengths, 315 and 365 nm (Figs. 3, 5, 6), can be represented by the following equation:

$$A(t) = \mathrm{A}\exp(-k_1 t) + \mathrm{B}\exp(-k_2 t) \tag{6}$$

where k_1 is associated with the appearance of the signal due to the G(-H)$^{\cdot}$ radicals and the decay of 2AP(-H)$^{\cdot}$ (recovery of 2AP), and k_2 is the observed rate constant of decay of the G(-H)$^{\cdot}$ radicals in subsequent reactions. The values of k_1 were found to be independent of the concentration of the duplexes in the 10–100 μM range. The rate constant k_1 is composed of two terms, k_a and k_{ag}, with $k_1=k_a+k_{ag}$. We denote the rate constant of guanine oxidation by k_{ag}, while the decay of the 2AP(-H)$^{\cdot}$ radical by other pathways (not involving electron transfer from G) is denoted by k_a. The rate constant k_a was measured using [2AP]A_{14} and [2AP]T_{14} duplexes; in these duplexes, the 2AP(-H)$^{\cdot}$ radicals cannot decay by electron transfer from G since there are no guanines (k_{ag}=0). The kinetic parameters determined by the best fits of Eq. 6 to the experimental transient absorption profiles are summarized in Table 2.

The effects of the number of intervening bases on the rate constant of guanine oxidation by the 2AP(-H)$^{\cdot}$ radicals is described by the following equation:

Table 2 Deuterium isotope effect on the rate constants of the oxidation of guanine and 8-oxoguanine by 2-aminopurine neutral radicals in double-stranded DNA duplexes (adapted from [11] and [13])

Sequence[a]	k_{ag} (10^3 s^{-1})		
	H_2O	D_2O	H_2O/D_2O
[2AP]$TGGT_{11}$	500±50		
[2AP]T_2GGT_{10}	10.3±1	5.9±0.6	1.7±0.2
[2AP]T_3GGT_9	3.3±0.3	2.2±0.2	1.5±0.2
[2AP]T_4GGT_8	0.9±0.1	n.d.[b]	n.d.
[2AP]T_2GT_{11}	9.9±1	5.7±0.6	1.7±0.2
[2AP]T_2GGGT_9	17.9±2	11.8±1.2	1.5±0.2
[2AP]A_2GGA_{10}	24.2±2	12.7±1.3	1.9±0.2
[2AP]A_4GGA_8	13.0±2	7.7±0.8	1.9±0.2
[2AP]A_6GGT_6	6.0±0.6	4.6±0.5	1.3±0.1
[2AP]A_8GGA_4	2.2±0.2	1.7±0.2	1.3±0.1
[2AP]T_2[8-oxo-dG]T_{10}	38±5	n.d.	n.d.
[2AP]T_4[8-oxo-dG]T_8	3.0±0.5	n.d.	n.d.

[a] Oligodeoxyribonucleotide sequences are written in the $5'\rightarrow3'$ direction
[b] n.d.=Not determined

$$k_{ag} = k_{ag}^0 \exp(-\beta r) \quad (7)$$

where β is an attenuation parameter and r is the distance between electron donor and acceptor residues. Plots of the linearized form of Eq. 7, using the available experimental data points for the duplexes with $n \geq 2$ intervening bases, are shown in Fig. 8. The [2AP]T_1GGT_{11} data were not included in this plot (as was done in [11]) because the contribution of the 2-μs component is rather small (~15%) in this case. However, a more detailed analysis allowed for the determination of the value of k_{ag} for the [2AP]T_4GGT_8 duplex, and this value has now been added to Fig. 7. Data [12] for the oxidation of 8-oxo-

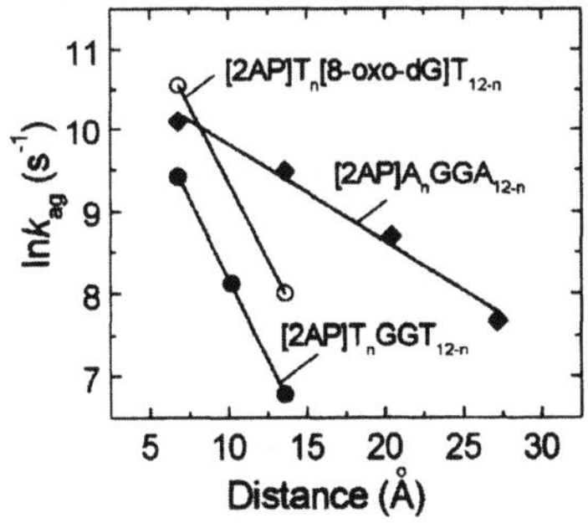

Fig. 8 The rate constants of proton-coupled electron transfer from guanine to 2AP(-H)$^{\cdot}$ radicals, k_{ag}, as a function of the number of bridging bases in double-stranded oligonucleotides (adapted from [11] and [13])

dG by the 2AP(-H)$^{\cdot}$ radical in the [2AP]T_n[8-oxo-dG]T_{13-n} duplexes (n=2 and 4) are also shown in Fig. 8. In these duplexes with T bridging bases, the oxidation of G or 8-oxo-dG by the 2AP(-H)$^{\cdot}$ radicals is the slow, but major pathway of the formation of the G(-H)$^{\cdot}$ radicals or 8-oxo- $G^{\cdot+}$/8-oxo-G(-H)$^{\cdot}$ radicals. The slopes of these plots are a measure of the parameter β. These values depend on the sequence and do not depend on the values of k_{ag}. For instance, the absolute values of k_{ag} for the oxidation of 8-oxo-dG are greater than those for the oxidation of guanine. Nevertheless, based on the two available data points for the sequences with 8-oxo-dG, the distance dependences for the duplexes with T sequences appear to be characterized by the same parameter β=0.4 Å^{-1} as the sequences with guanines. Note that a greater value of β=0.75 Å^{-1} was previously obtained using the data for the duplexes with one, two, and three T bridging residues [11]. In the duplexes with the A sequences, the distance dependence exhibits a more shallow dependence on the distance r since β=0.12 Å^{-1}. Thus, the A bridging sequences are more efficient than the T bridging sequences for promoting the oxidation of guanine not only by 2AP radical cations, but also by neutral 2AP radicals.

3.4
Proton-Coupled Electron Transfer at a Distance

The kinetic deuterium isotope effect on the rates of oxidation of guanine by 2AP radicals showed that this reaction, occurring at the level of free nucleosides, can be considered in terms of a proton-coupled electron transfer [14]. The solvent deuterium isotope effect was also used to probe the occurrence of proton-coupled electron transfer mechanisms in double-stranded DNA [13]. The values of k_{ag} in H_2O buffer solutions are larger than those in D_2O buffer solutions by factors of 1.3–1.7 (Fig. 9). The magnitude of the kinetic isotope effect, defined by the ratio $k_{ag}(H_2O)/k_{ag}$ (D_2O), decreases somewhat with an increase in the number of bridging bases (Table 2), as predicted by theoretical considerations of proton-coupled electron transfer reactions occurring at fixed distances [59]. In the DNA duplexes, the electron transfer from G to 2AP(-H)$^{\cdot}$ is coupled to a deprotonation of the radical cation, $G^{\cdot+}$, and a protonation of the anion, 2AP(-H)$^{-}$. In B-form DNA, the N2 sites of 2AP are accessible and the protonation of the 2AP(-H)$^{-}$ anion does not require any change in the hydrogen bonding configuration between the nucleic acid bases in the duplex.

The $dG^{\cdot+}$ radical cation is a Brønsted acid (pK_a=3.9) [48] and should rapidly deprotonate in neutral solutions of free nucleosides:

$$dG^{\bullet+} \rightleftharpoons dG(-H)^{\bullet} + H^{+} \tag{8}$$

Using this pK_a value and assuming that protonation of dG(-H)$^{\cdot}$ occurs with a diffusion-controlled rate constant (2×10^{10} M^{-1} s^{-1}) [48], we estimate that the deprotonation rate constant of $dG^{\cdot+}$ is ~2×10^{6} s^{-1}, i.e., the lifetime of the free $dG^{\cdot+}$ radical is about 500 ns at pH 7. In acid solutions the $G^{\cdot+}$ radical cations are relatively stable at room temperature and can be generated

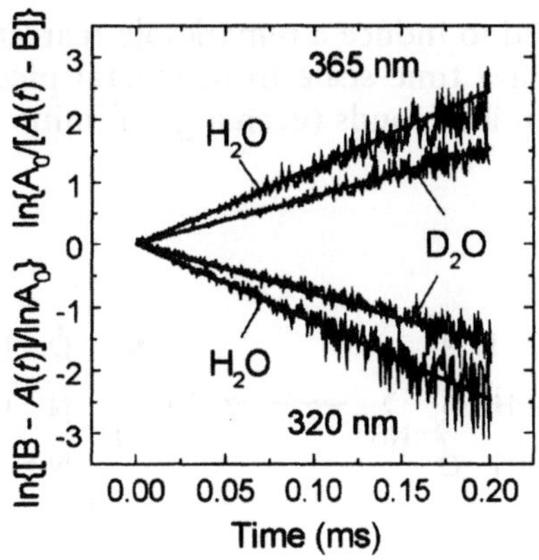

Fig. 9 Deuterium isotope effect on the kinetics of oxidation of G by 2AP(-H)$^{\cdot}$ radicals in the [2AP]T_2GGT_{10} duplex in oxygenated H_2O/D_2O buffer solutions (pH 7.0) [13]. The kinetic profiles (resolution of 0.5 μs/point) of the 2AP(-H)$^{\cdot}$ decay (365 nm) and the G(-H)$^{\cdot}$ formation (320 nm) were linearized according to a semilogarithmic form of Eq. 7. The *solid lines* are the best linear fits to the experimental data. Reprinted with permission from the J Phys Chem, Copyright (2001) American Chemical Society

by in situ photolysis methods at concentration ranges suitable for detection by conventional EPR techniques [79]. At pH<4, the $dG^{\cdot+}$ radical cation exhibits a spectrum consisting of 40 equidistant peaks with a separation of 0.068 mT centered at g=2.0038; this well-resolved hyperfine structure disappears at pH>5 due to the deprotonation of $dG^{\cdot+}$.

Deprotonation of the $dG^{\cdot+}$ radical cations in double-stranded DNA is evident from the EPR spectra of guanine radicals recorded in neutral solutions at room temperature [80, 81]. The EPR signal assigned to the neutral G(-H)$^{\cdot}$ radical derived from the deprotonation of $G^{\cdot+}$ shows the singlet with g~2.004 and half width at a half height ~0.8 mT. However, the deprotonation rate of $G^{\cdot+}$ cannot be estimated from the conventional EPR spectra, and further time-resolved EPR studies with laser pulse generation of the radicals are required to address this problem.

In double-stranded DNA, electron abstraction from the guanine radical cation can be associated with an extremely fast shift of the N1 proton to its Watson–Crick partner cytosine (Scheme 2a) [9]. The equilibrium constant for the protonation of C (pK_a=4.3) with the concomitant deprotonation of $G^{\cdot+}$ estimated from the pK values of the free nucleosides, is about 2.5 [49]. Within these constraints, the guanine radical should retain some radical cation character [82] and the complete deprotonation of $G^{\cdot+}$ would require a base pair opening event occurring on a millisecond timescale [74]. An alternative mechanism of $G^{\cdot+}$ deprotonation is the release of the N2 proton (Scheme 2b). This mechanism was experimentally established for 1-methylguanosine; conductometric results showed that in neutral solutions, the radical cation of this nucleoside rapidly deprotonates with the formation of the neutral radical [48]. Although the exact mechanism of the $G^{\cdot+}$ deprotonation in double-stranded DNA requires further clarification, electron abstraction from guanine is believed to induce a remarkable rearrangement of hydrogen bonding that occurs on a time scale from several picoseconds (N1 proton shift from $G^{\cdot+}$ to C) to milliseconds (exchange of imino protons via base pair opening).

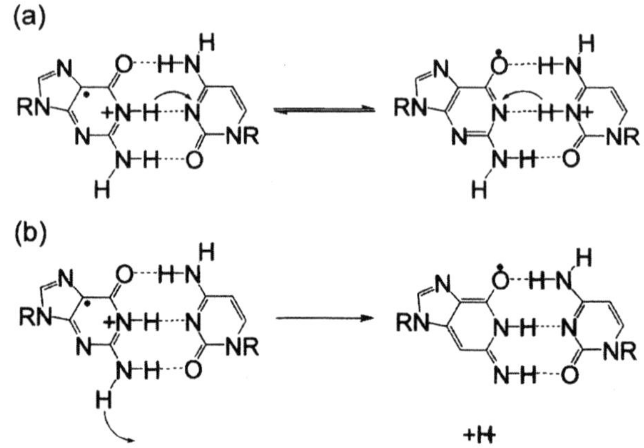

Scheme 2 Rearrangement of hydrogen bonding induced by hole localization at the G-C Watson–Crick base pair. **a** Proton shift from the $G^{\cdot+}$ radical cation to cytosine. **b** Dissociation of the N2 hydrogen bond in the $G^{\cdot+}$ radical cation (adapted from [9] and [48])

3.5 Oxidation of Guanine in the Sequence Context 5′-...G..., 5′-...GG..., and 5′-...GGG...

Oxidative modifications at guanine sites in DNA, detected after hot alkali treatment (alkali-labile lesions), exhibit a remarkable sequence selectivity [83–90]. The alkali-labile cleavage becomes more efficient when the number of tandem guanine bases is increased to two or three contiguous guanines. The cleavage at GG tandem sites is more efficient than at single G sites, and the cleavage at GGG triplets is more efficient than in GG doublets. These observed sequence effects correlate with the ionization potentials of the 5′-guanine, 7.51 eV (G), 7.28 eV (GG), and 7.07 eV (GGG) calculated for molecules in the gas phase [91]. However, the experimental rate constants for oxidation of G, GG, and GGG sequences by stilbene in the singlet excited state exhibit much smaller differences (1.0:1.7:1.5, respectively) [92]. The ratios of the k_{ag} values for oxidation of G, GG, and GGG by $2AP(-H)^{\cdot}$ in the following three sequences:

$$5' - [2AP]TTGTTTTTTTTTT - 3'$$
$$5' - [2AP]TTGGTTTTTTTTT - 3'$$
$$5' - [2AP]TTGGGTTTTTTTT - 3' \qquad (d)$$

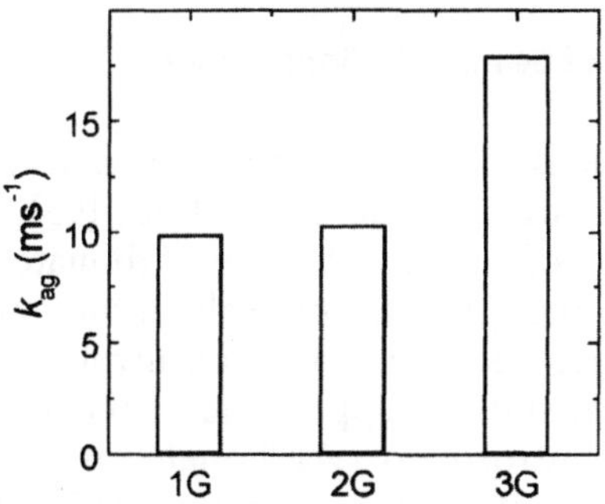

Fig. 10 The rate constants of proton-coupled electron transfer from guanine to 2AP(-H)$^{\cdot}$ radicals, k_{ag}, in the [2AP]$T_2(G)_nT_{10}$ duplexes

are also very close to one another (Fig. 10). Of course, which of the two or three G bases is oxidized cannot be resolved, but it is most likely the 5′-G in each case.

Figure 10 shows that the values of k_{ag} for the oxidation of G and GG are close to one another, and are smaller by a factor of only ~1.7 than the value of k_{ag} for the oxidation of guanine in the GGG sequence. Estimates by Lewis et al. [89] have shown that even small differences in the rate constants can provide modest selectivities for alkali-labile strand cleavage observed in a number of experimental studies [83–90].

4
The Guanine Radical as a Key Intermediate in the Formation of Oxidatively Modified Guanine Bases in DNA

In neutral aqueous solutions, the ultimate product of one-electron abstraction from guanine is the guanine neutral radical. In DNA, this radical is formed via the deprotonation of the guanine radical cation arising, e.g., from hole localization or directly via proton-coupled electron transfer from guanine to an appropriate electron acceptor. The G(-H)$^{\cdot}$ radicals do not exhibit observable reactivities with molecular oxygen ($k \leq 10^2$ M^{-1} s^{-1}) [93]. Steenken et al. have concluded that in double-stranded DNA direct hydrogen atom abstraction from 2′-deoxyribose by G(-H)$^{\cdot}$ radical is very unlikely due to steric hindrance effects and a small thermodynamic driving force [94]. The EPR studies performed in neutral aqueous solutions at room temperature have shown that, in the absence of specific reactive molecules, the lifetime of the G(-H)$^{\cdot}$ radical in double-stranded DNA is as long as ~5 s [80]. Therefore, the fates of G(-H)$^{\cdot}$ radicals are mostly determined by the presence of other reactive species and radicals. Thus, the G(-H)$^{\cdot}$ radical can be a key precursor of diverse guanine lesions in DNA. In the next section we begin from a discussion of the site-selective generation of the G(-H)$^{\cdot}$ radical in DNA, and then continue with a discussion of the reaction pathways of this guanine radical.

4.1 Site-Selective Generation of Guanine Radicals in DNA

Guanine is the most easily oxidizable natural nucleic acid base [8] and many oxidants can selectively oxidize guanine in DNA [95]. Here, we focus on the site-selective oxidation of guanine by the carbonate radical anion, $CO_3^{\cdot-}$, one of the important emerging free radicals in biological systems [96]. The mechanism of $CO_3^{\cdot-}$ generation in vivo can involve one-electron oxidation of HCO_3^- at the active site of copper-zinc superoxide dismutase [97, 98], and homolysis of the nitrosoperoxycarbonate anion ($ONOOCO_2^-$) formed by the reaction of peroxynitrite with carbon dioxide [99–102].

The $CO_3^{\cdot-}$ radical anion is a strong one-electron oxidant (E_7~1.7 V vs NHE [15]) that oxidizes appropriate electron donors via electron transfer mechanisms [103]. Detailed pulse radiolysis studies have shown that carbonate radicals can rapidly abstract electrons from aromatic amino acids (tyrosine and tryptophan). However, reactions of $CO_3^{\cdot-}$ with S-containing methionine and cysteine are less efficient [104–106]. Hydrogen atom abstraction by carbonate radicals is generally very slow [103] and their reactivities with other amino acids are negligible [104–106].

We have investigated the reactions of the $CO_3^{\cdot-}$ radicals with double-stranded DNA by laser flash photolysis techniques [15]. In these time-resolved experiments, the $CO_3^{\cdot-}$ radicals were generated by one-electron oxidation of HCO_3^- by sulfate radical anions, $SO_4^{\cdot-}$; the latter were derived from the photodissociation of persulfate anions, $S_2O_8^{2-}$ initiated by 308-nm XeCl excimer laser pulse excitation. In air-equilibrated buffer solution containing the self-complementary oligonucleotide duplex d(AACGCGAATTCGCGTT), $S_2O_8^{2-}$, and an excess of HCO_3^-, the decay of the $CO_3^{\cdot-}$ radical anion absorption band at 600 nm is associated with the concomitant formation of the characteristic narrow absorption band of the G(-H)$^\cdot$ radicals near 310 nm.

The selectivity of the guanine oxidation by the $CO_3^{\cdot-}$ radicals was revealed by treatment with hot piperidine solutions, which gives rise to strand cleavage at the damaged DNA sites [107]. The cleaved fragments thus formed can be visualized by high-resolution gel electrophoresis. Typical results are depicted in the gel autoradiograph in Fig. 11a. It is evident that the sites of cleavage correlate well with the cleavage patterns obtained by the Maxam-Gilbert sequencing reaction (lane G), clearly indicating that strand cleavage occurs predominantly at the four different guanines in the duplex. The strand cleavage efficiencies (Fig. 11b) are lower at the two central guanines G^2 and G^3 than at the outer guanines G^1 and G^4. These differences in reactivities correlate roughly with the imino proton exchange rates, and thus the base pair opening rates, in the different G:C base pairs in the 12-mer d($CG^4CG^3AATTCG^2CG^1$) studied by Patel and coworkers [108]. The efficiencies of strand cleavage at sites other than guanine are significantly smaller (Fig. 11c). These results clearly indicate that the $CO_3^{\cdot-}$ radical is a site-selective oxidizing agent of guanines in double-stranded DNA. Following irradiation, the predominant cleavage at the guanine sites, induced by treatment with hot piperidine, is consistent with the spectroscopic laser photolysis

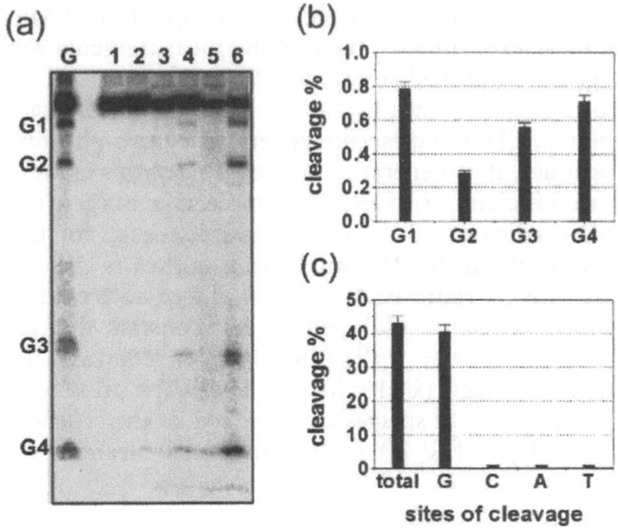

Fig. 11a–c Gel electrophoresis patterns of irradiated, hot piperidine-treated oligonucleotide strands [15]. Autoradiograph of a denaturating gel (7 M urea, 20% polyacrylamide gel) showing the cleavage patterns induced by 308-nm excimer laser pulse excitation (20 mJ pulse^{-1} cm^2) of the 5′-^{32}P-end-labeled d(AACG^{4}CG^{3}AATTCG^{2}CG^{1}TT) strands in the duplex form (10 μM) in air-equilibrated buffer solutions (pH 7.5) containing 10 mM $S_2O_8^{2-}$ and 300 mM HCO_3^-. After the irradiation, the oligonucleotide solutions were treated with hot piperidine (1 M, 90 °C) for 30 min and loaded onto a polyacrylamide gel. **a** Lane G: Maxam–Gilbert G sequencing reaction. Lane 1: unirradiated duplex (without piperidine treatment); lane 2: unirradiated duplex (after hot piperidine treatment); lanes 3 and 5 (integrated dosage received by the sample, 0.5 and 4 J cm^{-2}, respectively): irradiated duplex (without piperidine treatment); lanes 4 and 6 (integrated dosage received by the sample, 0.5 and 4 J cm^{-2}, respectively): irradiated duplex (after hot piperidine treatment). **b** Histogram obtained by scanning the autoradiogram (lane 4, 0.5 J cm^{-2}): hot piperidine cleavage patterns at the different guanine sites G^1, G^2, G^3, and G^4 in the irradiated, self-complementary 5′-^{32}P-end-labeled d(AACG^{4}C-G^{3}AATTCG^{2}CG^{1}TT) duplex. **c** Total fraction of strands cleaved at the dosage of 4 J cm^{-2}, and the fraction of fragments cleaved at all G, C, A, and T sites in the self-complementary duplex (from lane 6). Reprinted with permission from the J Biol Chem, Copyright (2001) American Society for Biochemistry and Molecular Biology

data that indicate the selective oxidation of guanines by $CO_3^{\cdot-}$ radicals. The selective reactivity of the $CO_3^{\cdot-}$ radicals with guanine, rather than with any of the other three DNA bases, is a consequence of the thermodynamic and kinetic characteristics of these electron donor/acceptor reactions [15].

In the double-stranded oligonucleotide, the lifetimes of the G(-H)$^{\cdot}$ radicals are significantly longer than the lifetimes of either dGMP(-H)$^{\cdot}$ or dG(-H)$^{\cdot}$ mononucleoside radicals in solution [10, 50, 109]. Furthermore, the lifetimes of the G(-H)$^{\cdot}$ radicals in the self-complementary d(AACGC-GAATTCGCGTT) duplex is of the order of seconds [15], i.e., orders of magnitude longer than in oligonucleotide duplexes in which these radicals are generated by one-electron transfer reactions from G to 2AP(-H)˙ radicals [11]. In all of these experiments, the lifetime measurements were conducted in air- or oxygen-equilibrated solutions. Thus, differences in the decay rates due to reactions of G(-H)˙ with O_2 can be ruled out. In the case of the 2AP-modified duplexes, 2AP radicals were generated by two-photon photoionization of 2AP residues, thus generating hydrated electrons as well. The hydrated electrons are efficiently scavenged by molecular oxygen resulting in the formation of $O_2^{\cdot-}$ radicals [51]. Hence, a possible reason for the faster decay of the G(-H)˙ radicals in the 2AP-modified duplexes is the fast reaction of G(-H)˙ radicals with $O_2^{\cdot-}$ radicals. At the level of free nucleosides, the reaction of dG(-H)˙ with $O_2^{\cdot-}$ occurs rapidly with a rate constant that is nearly diffusion controlled ($1–3\times10^9$ M^{-1} s^{-1}) [109, 110]. In contrast, the reaction of G(-H)˙ with O_2 is very slow ($k\leq10^2$ M^{-1} s^{-1}) [93]. The presence or absence of $O_2^{\cdot-}$ or other reactive radical species could be one of the crucial factors which determines the lifetime of G(-H)˙ radicals in double-stranded DNA.

4.2
Bimolecular Reactions of Guanine Radicals with Other Radical Species

The fates of the G(-H)˙ radicals in DNA are mostly determined by reactions with other substrates. Here, we consider the reactions of the G(-H)˙ radicals with types of free radicals that are generated in vivo under conditions of oxidative stress. One of these radicals is the nitrogen dioxide radical, ˙NO_2. This radical can be generated in vivo by the oxidation of nitrite, NO_2^-, a process that can be mediated by myeloperoxidase [111, 112] as well as by other cellular oxidants [113, 114]. An alternative pathway of the generation of ˙NO_2 is the homolysis of peroxynitrite [102, 115] or nitrosoperoxycarbonate formed by the reaction of peroxynitrite with carbon dioxide [99–101]. The redox potential, E^0(˙NO_2/NO_2^-)=1.04 V vs NHE [116] is less than that of guanine, E_7[G(-H)˙/G]=1.29 V vs NHE [8]. Pulse radiolysis [117] and laser flash photolysis [109] experiments have shown that, in agreement with these redox potentials, ˙NO_2 radicals do not react with intact DNA. However, ˙NO_2 radicals can oxidize 8-oxo-dG that has a lower redox potential (E_7=0.74 vs NHE [56]) than any of the normal nucleobases [109].

Although the ˙NO_2 radicals do not react with the intact DNA bases, they readily combine with the guanine radicals. We used this bimolecular combination reaction for the generation of diverse nitroguanine products [118]. The overall rationale of our approach is outlined in Scheme 3. The G(-H)˙ radicals in a single-stranded oligonucleotide are produced by the site-selective oxidation of guanine by $CO_3^{\cdot-}$ radical anions [15]. The combination reaction of G(-H)˙ with ˙NO_2 occurs via two competitive pathways involving the addition of ˙NO_2 to C8 with the formation of the 8-nitro-dG adduct, or to C5 with the formation of an unstable adduct which spontaneously collapses to the stable 5-guanidino-4-nitroimidazole adduct. The site-specific 8-nitro-dG and 5-guanidino-4-nitroimidazole adducts thus formed were isolated by reversed-phase HPLC (Fig. 12), and identified by mass spectrometry techniques [118].

The combination of $^{\cdot}NO_2$ radicals with diverse radicals is a very rapid reaction occurring with rates close to the diffusion-controlled limit [103]. Nitrogen and/or oxygen atoms of $^{\cdot}NO_2$ radicals can participate in the formation of chemical bonds with the target radical, because the unpaired electron

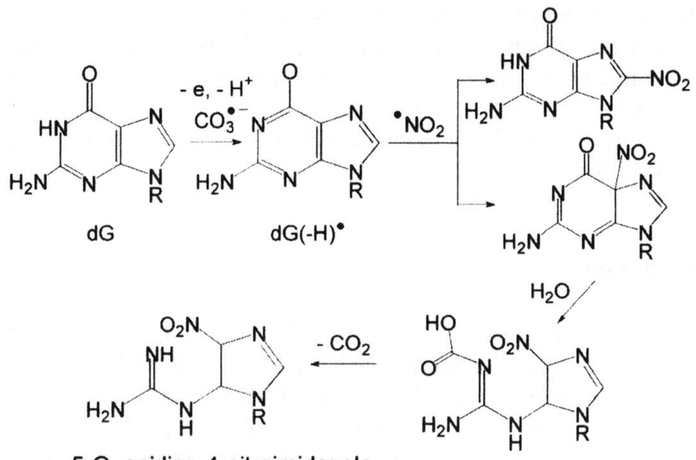

Scheme 3 Nitration of guanine in single-stranded DNA via combination of the photochemically generated $G(-H)^{\cdot}$ and $^{\cdot}NO_2$ radicals

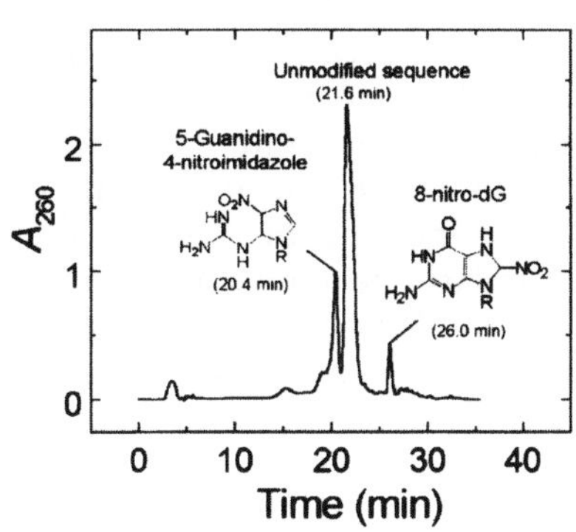

Fig. 12 Reversed-phase HPLC elution profile of the nitration products photochemically generated by a 20-s train of 308-nm XeCl excimer laser pulses (15 mJ pulse^{-1} cm^{-2}, 10 pulse s^{-1}) in air-equilibrated 20 mM phosphate buffer solution (pH 7.5) containing 0.1 mM of the single-stranded oligonucleotide, 5′-d(CCATCGCTACC), 300 mM $NaHCO_3$, 1 mM $NaNO_2$, and 25 mM $Na_2S_2O_8$. HPLC elution conditions: 11–20% linear gradient of acetonitrile in 50 mM triethylammonium acetate (pH 7) for 60 min at a flow rate of 1 mL min^{-1}. The unmodified sequence 5′-d(CCATCGCTACC) is eluted at 21.6 min, and the nitrated sequences containing 5-guanidino-4-nitroimidazole and 8-nitro-dG lesions elute at 20.4 and 26.0 min, respectively. Reprinted with permission from Chem Res Toxicol, Copyright (2001) American Chemical Society

is delocalized on both N and O atoms [119]. In the case of organic radicals, their combination with $^{\cdot}NO_2$ can result in the formation of nitro compounds and nitroso oxy derivatives. Typically, the nitro products are more stable than the nitroso oxy compounds and can be isolated from the reaction mixture. For instance, combination of the $^{\cdot}NO_2$ and tyrosine radicals occurs with a rate constant of $\sim 3\times10^9\ M^{-1}\ s^{-1}$ and results in the formation of 3-nitrotyrosine [117]. The combination of the $^{\cdot}NO_2$ and G(-H)$^{\cdot}$ radicals also results in the formation of the relatively stable nitro end products. The photochemical nitration of single-stranded oligonucleotide generates the 8-nitro-dG and 5-guanidino-4-nitroimidazole products in the ratio of ~1:2 (Fig. 12). This ratio can be attributed to the interplay of diverse factors, such as the delocalization of unpaired electrons on C8 and C5 sites, and the stabilities of the C8 and C5 nitro end products.

5 Concluding Remarks

The selective two-photon ionization of 2AP residues by intense 308-nm excimer laser excitation provides a novel method of site-selective injection of holes into double-stranded DNA. Localized holes, the 2AP radical cations, are very short-lived and transform into more stable 2AP neutral radicals which, like the radical cation precursors, are also strong one-electron oxidants. Transient absorption measurements within the nanosecond to millisecond time domains revealed that the kinetics of guanine oxidation at a distance are highly heterogeneous. The heterogeneous character of the kinetics of guanine radical formation is, in part, associated with the existence of two oxidants that have very different lifetimes in DNA, the 2AP radical cation and the 2AP neutral radical. Spectroscopic kinetic measurements showed that guanine oxidation at a distance, in DNA duplexes with adenine bridging bases between the guanine electron donor and 2AP radical acceptor on the same strand, is significantly faster than in the case of bridging thymidines. The oxidation of guanine in the sequence context 5′-...G..., 5′-...GG..., and 5′-...GGG... occurs with similar rate constants. Using ultrafast laser kinetic spectroscopy techniques, similar observations have been reported in DNA hairpins containing stilbene chromophores that oxidize guanines by one-electron transfer mechanisms upon photoexcitation [78, 92, 120, 121].

The solvent kinetic isotope effects on the rates of guanine oxidation by 2AP neutral radicals indicate that electron transfer reactions from guanine to 2AP radicals occurring at a distance in DNA duplexes are coupled with a deprotonation/protonation mechanism. These proton-coupled electron transfer reactions at a distance generate neutral guanine radicals. Due to the low reactivity toward molecular oxygen and 2′-deoxyribose moieties, guanine radicals are very long-lived (~seconds) in double-stranded DNA. Trapping reactions of guanine radicals with biologically relevant free radicals lead to the formation of diverse oxidatively modified guanine bases.

Acknowledgments The research described here has been supported by the National Science Foundation, Grant CHE-9700429, by the National Institutes of Health, Grant 5-R01-ES011589, and by a grant from the Kresge Foundation.

References

1. Beckman KB, Ames BN (1997) J Biol Chem 272:19633
2. Grisham MB, Jourd'heuil D, Wink DA (2000) Aliment Pharmacol Ther 14 Suppl 1:3
3. Cadet J, Bourdat AG, D'Ham C, Duarte V, Gasparutto D, Romieu A, Ravanat JL (2000) Mutat Res 462:121
4. Nunez ME, Hall DB, Barton JK (1999) Chem Biol 6:85
5. Schuster GB (2000) Acc Chem Res 33:253
6. Giese B (2002) Annu Rev Biochem 71:51
7. Hall DB, Holmlin RE, Barton JK (1996) Nature 382:731
8. Steenken S, Jovanovic SV (1997) J Am Chem Soc 119:617
9. Steenken S (1997) Biol Chem 378:1293
10. Shafirovich V, Dourandin A, Huang W, Luneva NP, Geacintov NE (1999) J Phys Chem B 103:10924
11. Shafirovich V, Dourandin A, Huang W, Luneva NP, Geacintov NE (2000) Phys Chem Chem Phys 2:4399
12. Shafirovich V, Cadet J, Gasparutto D, Dourandin A, Huang W, Geacintov NE (2001) J Phys Chem B 105:586
13. Shafirovich V, Dourandin A, Geacintov NE (2001) J Phys Chem B 105:8431
14. Shafirovich V, Dourandin A, Luneva NP, Geacintov NE (2000) J Phys Chem B 104:137
15. Shafirovich V, Dourandin A, Huang W, Geacintov NE (2001) J Biol Chem 276:24621
16. Ronen A (1980) Mutat Res 75:1
17. Nordlund TM, Andersson S, Nilsson L, Rigler R, Gräslund A, McLaughlin LW (1989) Biochemistry 28:9095
18. Law SW, Eritja R, Goodman MF, Breslauer KJ (1996) Biochemistry 35:12329
19. Sowers LC, Fazakerley GV, Eritja R, Kaplan BE, Goodman MF (1986) Proc Natl Acad Sci USA 83:5434
20. Fagan PA, Fabrega C, Eritja R, Goodman MF, Wemmer DE (1996) Biochemistry 35:4026
21. Sowers LC, Boulard Y, Fazakerley GV (2000) Biochemistry 39:7613
22. Callis PR (1983) Annu Rev Phys Chem 34:329
23. Holmén A, Nordén B, Albinsson B (1997) J Am Chem Soc 119:3114
24. Rigler R, Claesens F (1986) In: Van Knippenberg PH, Hilbers CW (eds) Structure and dynamics of RNA. Plenum, New York, p 45
25. Guest CR, Hochstrasser RA, Sowers LC, Millar DP (1991) Biochemistry 30:3271
26. Hochstrasser RA, Carver TE, Sowers LC, Millar DP (1994) Biochemistry 33:11971
27. Kelley SO, Barton JK (1999) Science 283:375
28. Wan C, Fiebig T, Schiemann O, Barton JK, Zewail AH (2000) Proc Natl Acad Sci USA 97:14052
29. Larsen OFA, Van Stokkum IHM, Gobets B, Van Grondelle R, Van Amerongen H (2001) Biophys J 81:1115
30. Bernards AS, Miller JK, Bao KK, Wong I (2002) J Biol Chem 277:20960
31. Ying J, Bradley RK, Jones LB, Reddy MS, Colbert DT, Smalley RE, Hardin SH (1999) Biochemistry 38:16461
32. Mandal SS, Fidalgo da Silva E, Reha-Krantz LJ (2002) Biochemistry 41:4399
33. Dunlap CA, Tsai MD (2002) Biochemistry 41:11226
34. Stivers JT (1998) Nucleic Acids Res 26:3837
35. Allan BW, Reich NO (1996) Biochemistry 35:14757
36. Holz B, Klimasauskas S, Serva S, Weinhold E (1998) Nucleic Acids Res 26:1076

37. Malygin EG, Evdokimov AA, Zinoviev VV, Ovechkina LG, Lindstrom WM, Reich NO, Schlagman SL, Hattman S (2001) Nucleic Acids Res 29:2361
38. Larsen OFA, Van Stokkum IHM, de Weerd FL, Vengreis M, Aravinda Kumar CT, Van Grondelle R, Geacintov NE, Van Amerongen H (2003) Phys Chem Chem Phys (accepted)
39. Jean JM, Hall KB (2001) Proc Natl Acad Sci USA 98:37
40. Pal SK, Peon J, Zewail AH (2002) Chem Phys Lett 363:57
41. Hart EJ, Boag JW (1962) J Am Chem Soc 84:4090
42. Candeias LP, Wolf P, O'Neill P, Steenken S (1992) J Phys Chem 96:10302
43. Nikogosyan DN, Gurzadyan GG (1984) Laser Chem 4:297
44. Candeias LP, O'Neill P, Jones GDD, Steenken S (1992) Int J Radiat Biol 61:15
45. Melvin T, Botchway SW, Parker AW, O'Neill PJ (1995) J Chem Soc Chem Commun 653
46. Melvin T, Botchway SW, Parker AW, O'Neill P (1996) J Am Chem Soc 118:10031
47. Candeias LP, Steenken S (1992) J Am Chem Soc 114:699
48. Candeias LP, Steenken S (1989) J Am Chem Soc 111:1094
49. Steenken S (1992) Free Radic Res Commun 16:349
50. Shafirovich V, Dourandin A, Luneva NP, Geacintov NE (2000) J Chem Soc Perkin Trans 2:271
51. Buxton GV, Greenstock CL, Helman WP, Ross AB (1988) J Phys Chem Ref Data 17:513
52. Beckman KB, Ames BN (1999) Mutat Res 424:51
53. Baik SC, Youn HS, Chung MH, Lee WK, Cho MJ, Ko GH, Park CK, Kasai H, Rhee KH (1996) Cancer Res 56:1279
54. Musarrat J, Arezina-Wilson J, Wani AA (1996) Eur J Cancer 32A:1209
55. Li D, Zhang W, Sahin AA, Hittelman WN (1999) Cancer Detect Prev 23:454
56. Steenken S, Jovanovic SV, Bietti M, Bernhard K (2000) J Am Chem Soc 122:2373
57. Cukier RI, Nocera DG (1998) Annu Rev Phys Chem 49:337
58. Cukier RI (2002) J Phys Chem B 106:1746
59. Decornez H, Hammes-Schiffer S (2000) J Phys Chem A 104:9370
60. Carra C, Iordanova N, Hammes-Schiffer S (2002) J Phys Chem B 106:8415
61. Kuzmin VA, Dourandin A, Shafirovich V, Geacintov NE (2000) Phys Chem Chem Phys 2:1531
62. O'Connor D, Shafirovich VY, Geacintov NE (1994) J Phys Chem 98:9831
63. Shafirovich VY, Courtney SH, Ya N, Geacintov NE (1995) J Am Chem Soc 117:4920
64. Weatherly SC, Yang IV, Thorp HH (2001) J Am Chem Soc 123:1236
65. Cai Z, Li X, Sevilla MD (2002) J Phys Chem B 106:2755
66. Razskazovskii Y, Roginskaya M, Sevilla MD (1998) Radiat Res 149:422
67. Taylor J, Eliezer I, Sevilla MD (2001) J Phys Chem B 105:1614
68. Kawai K, Wata Y, Ichinose N, Majima T (2000) Angew Chem Int Ed 39:4327
69. Kawai K, Wata Y, Hara M, Tojo S, Majima T (2002) J Am Chem Soc 124:3586
70. Nikogosyan DN, Angelov DA, Oraevsky AA (1982) Photochem Photobiol 35:627
71. Nikogosyan DN (1990) Int J Radiat Biol 57:233
72. Nikogosyan DN, Letokhov VS (1983) Riv Nuovo Cimento 6:1
73. Gueron M, Kochoyan M, Leroy JL (1987) Nature 328:89
74. Kochoyan M, Leroy JL, Gueron M (1987) J Mol Biol 196:599
75. Leroy JL, Charretier E, Kochoyan M, Gueron M (1988) Biochemistry 27:8894
76. Jortner J, Bixon M, Langenbacher T, Michel-Beyerle ME (1998) Proc Natl Acad Sci USA 95:12759
77. Bixon M, Giese B, Wessely S, Langenbacher T, Michel-Beyerle ME, Jortner J (1999) Proc Natl Acad Sci USA 96:11713
78. Lewis FD, Zuo X, Liu J, Hayes RT, Wasielewski MR (2002) J Am Chem Soc 124:4568
79. Bachler V, Hildenbrand K (1992) Radiat Phys Chem 40:59
80. Hildenbrand K, Schulte-Frohlinde D (1990) Free Radic Res Commun 11:195
81. Schiemann O, Turro NJ, Barton JK (2000) J Phys Chem B 104:7214
82. Hutter M, Clark T (1996) J Am Chem Soc 118:7574
83. Kovalsky OI, Panyutin IG, Budowsky EI (1990) Photochem Photobiol 52:509

84. Ito K, Inoue S, Yamamoto K, Kawanishi S (1993) J Biol Chem 268:13221
85. Saito I, Takayama M, Sugiyama H, Nakatani K, Tsuchida A, Yamamoto M (1995) J Am Chem Soc 117:6406
86. Muller JG, Hickerson RP, Perez RJ, Burrows CJ (1997) J Am Chem Soc 119:1501
87. Saito I, Nakamura T, Nakatani K, Yoshioka Y, Yamaguchi K, Sugiyama H (1998) J Am Chem Soc 120:12686
88. Yoshioka Y, Kitagawa Y, Takano Y, Yamaguchi K, Nakamura T, Saito I (1999) J Am Chem Soc 121:8712
89. Hickerson RP, Prat F, Muller JG, Foote CS, Burrows CJ (1999) J Am Chem Soc 121:9423
90. Nakatani K, Dohno C, Saito I (2000) J Am Chem Soc 122:5893
91. Sugiyama H, Saito I (1996) J Am Chem Soc 118:7063
92. Lewis FD, Liu X, Liu J, Hayes RT, Wasielewski MR (2000) J Am Chem Soc 122:12037
93. Al-Sheikhly M (1994) Radiat Phys Chem 44:297
94. Steenken S, Jovanovic SV, Candeias LP, Reynisson J (2001) Chem Eur J 7:2829
95. Kawanishi S, Hiraku Y, Oikawa S (2001) Mutat Res 488:65
96. Augusto O, Bonini MG, Amanso AM, Linares E, Santos CC, De Menezes SL (2002) Free Radic Biol Med 32:841
97. Goss SP, Singh RJ, Kalyanaraman B (1999) J Biol Chem 274:28233
98. Zhang H, Joseph J, Felix C, Kalyanaraman B (2000) J Biol Chem 275:14038
99. Lymar SV, Hurst JK (1998) J Am Chem Soc 37:294
100. Goldstein S, Czapski G (1999) J Am Chem Soc 121:2444
101. Bonini MG, Radi R, Ferrer-Sueta G, Ferreira AM, Augusto O (1999) J Biol Chem 274:10802
102. Hodges GR, Ingold KU (1999) J Am Chem Soc 121:10695
103. Neta P, Huie RE, Ross AB (1988) J Phys Chem Ref Data 17:1027
104. Adams GE, Aldrich JE, Bisby RH, Cundall RB, Redpath JL, Willson RL (1972) Radiat Res 49:278
105. Chen S-N, Hoffman MZ (1973) Radiat Res 56:40
106. Baverstock KF, Cundall RB, Adams GE, Redpath JL (1974) Int J Radiat Biol 26:39
107. Burrows CJ, Muller JG (1998) Chem Rev 98:1109
108. Patel DJ, Pardi A, Itakura K (1982) Science 216:581
109. Shafirovich V, Cadet J, Gasparutto D, Dourandin A, Geacintov NE (2001) Chem Res Toxicol 14:233
110. Candeias LP, Steenken S (2000) Chem Eur J 6:475
111. Van der Vliet A, Eiserich JP, Halliwell B, Cross CE (1997) J Biol Chem 272:7617
112. Byun J, Mueller DM, Fabjan JS, Heinecke JW (1999) FEBS Lett 455:243
113. Eiserich JP, Cross CE, Jones AD, Halliwell B, van der Vliet A (1996) J Biol Chem 271:19199
114. Jiang Q, Hurst JK (1997) J Biol Chem 272:32767
115. Gerasimov OV, Lymar SV (1999) Inorg Chem 38:4317
116. Stanbury DM (1989) Adv Inorg Chem 33:69
117. Prütz WA, Monig H, Butler J, Land EJ (1985) Arch Biochem Biophys 243:125
118. Shafirovich V, Mock S, Kolbanovskiy A, Geacintov NE (2002) Chem Res Toxicol 15:591
119. Behar D, Fessenden RW (1972) J Phys Chem 76:1710
120. Lewis FD, Liu X, Liu J, Miller S, Hayes RT, Wasielewski MR (2000) Nature 406:51
121. Lewis FD, Wu T, Liu X, Letsinger RL, Greenfield SR, Miller SE, Wasielewski MR (2000) J Am Chem Soc 122:2889

Top Curr Chem (2004) 237:159–181
DOI 10.1007/b94476

Electrocatalytic DNA Oxidation

H. Holden Thorp

Department of Chemistry, University of North Carolina, Chapel Hill, NC 27599-3290, USA
E-mail: holden@unc.edu

1 Introduction

The redox reactions of the guanine nucleobase have received considerable attention because of the importance of these redox reactions in aging and metabolism [1], the potential utility of these oxidations in analytical techniques for detection of nucleic acids [2–5], and for use as donors in studies of electron-transfer reactions along the DNA double helix [6–8]. Electrochemistry is a powerful technique for studying the redox reactions of guanine because modern digital methods provide detailed information on kinetics and thermodynamics [9, 10], and because the small depth of the diffusion layer requires small quantities of material for analysis [11]. Thus, electrochemistry offers the possibility of performing detailed mechanistic studies on quantities of material much smaller than those required for optical methods, thereby allowing routine study of numerous sequences prepared using solid-phase synthesis or the techniques of molecular biology [12].

The observation of currents attributable to the faradaic electrochemistry of nucleic acids was pioneered by Palecek and coworkers who studied DNA adsorbed on mercury or carbon electrodes [13]. The signals detected by Palecek were attributable to oxidation of the purines, which produced signals indicative of irreversible processes involving adsorbed bases. These reactions were used as a basis for electrochemical analysis of DNA. Kuhr and coworkers later showed that similar strategies could be developed for analysis of nucleic acids via oxidation of sugars at copper electrodes [14–16].

This article will address research done in our group where complexes similar to $Ru(bpy)_3^{2+}$ (bpy=2,2′-bipyridine) are used to oxidize guanine and related nucleobases by a single electron upon generation of the Ru(III) state of the complex [17]. The reaction is readily monitored as a catalytic current in the oxidation of the complex to $Ru(bpy)_3^{3+}$, and we have developed detailed methods for obtaining kinetic and thermodynamic information on the guanine–Ru(III) reaction from these signals [18, 19]. This information can be supplemented with parallel studies of the same reactions by following the optical absorption of the metal complex using stopped-flow absorption spectrophotometry or following the fate of the oxidized guanine by gel electrophoresis [12, 18]. This latter method provides sequence-specific information that reveals the effects of different DNA sequences and structures on the reactivity of the nucleobase. Although we have characterized the surface chemistry [20–24] and electrochemical mechanisms [25–27] of these reactions in detail, the discussion will be limited to those aspects that pertain to the general issues of long-range electron transfer, which is generally the subject of this volume. The electrochemical reactions we have studied occur mostly at short range, but many of the findings are relevant to questions surrounding the long-range systems. The final section will relate these findings to the issues in long-range electron transfer.

2 Observation of Electrocatalytic Guanine Oxidation

The electrochemical approach discussed here relies on a number of special properties of indium tin-oxide (ITO) electrodes, which had been used in particular for spectroelectrochemistry since ITO is optically transparent and can be fabricated on glass [28, 29]. The first important attribute of ITO is the ability to access potentials up to about 1.4 V (all potentials versus SSCE) in neutral solution [29]. Second, ITO electrodes do not adsorb DNA appreciably [30], which could be anticipated from the ability of metal oxides to adsorb cationic proteins [31]; polyanionic nucleic acids were therefore not expected to adsorb. This property makes ITO quite different from carbon, which allows access to relatively high potentials but strongly adsorbs DNA [32]. Third, the direct oxidation of guanine at ITO is extremely slow, even

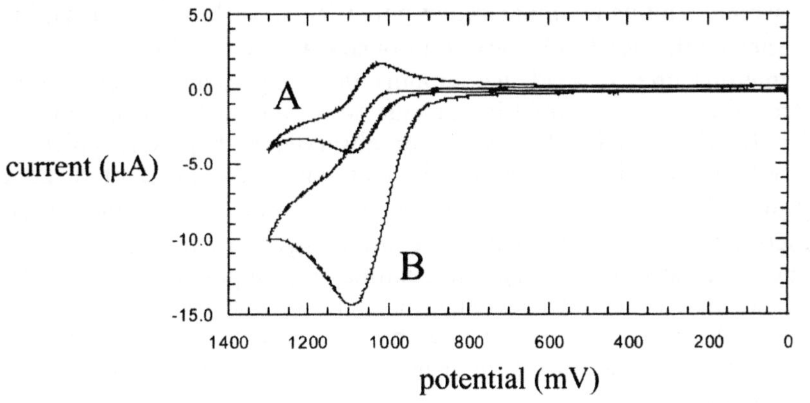

Fig. 1a,b Cyclic voltammograms at ITO working electrodes of $Ru(bpy)_3^{2+}$ without (**a**) and with (**b**) herring testes DNA at 800 mM added salt. The scan rate is 100 mV/s, and all potentials are versus Ag/AgCl

when DNA is intentionally adsorbed on the electrode [22, 23], so signals from direct guanine oxidation are generally not significant.

These described properties of ITO electrodes allow for development of a system for studying electrocatalytic DNA oxidation. Figure 1 shows the cyclic voltammogram of $Ru(bpy)_3^{2+}$ at ITO in phosphate buffer at pH 7 with 800 mM added salt (solid). As seen in Fig. 1a, quasi-reversible voltammograms for $Ru(bpy)_3^{2+}$ are readily detected at conventional scan rates. Figure 1b shows the signal obtained upon the addition of herring testes DNA. The wave detected is indicative of a scheme where the electrogenerated Ru(III) is reduced to Ru(II) by DNA, setting up an electrocatalytic cycle [17, 33]:

$$Ru(bpy)_3^{2+} \rightarrow Ru(bpy)_3^{3+} + e^- \quad (1)$$

$$Ru(bpy)_3^{3+} + \text{guanine} \rightarrow Ru(bpy)_3^{2+} + \text{guanine}_{ox} \quad (2)$$

where guanine_{ox} has been oxidized by a single electron compared to native guanine. The likely fate of the oxidized guanine will be discussed in more detail below. Thus, when the electrode is swept through the potential for the Ru(III/II) oxidation, the electrogenerated $Ru(bpy)_3^{3+}$ undergoes a thermal reaction with guanine to regenerate $Ru(bpy)_3^{2+}$, which is oxidized again by the electrode. The additional current in the electrocatalytic wave therefore arises from the reoxidation of the $Ru(bpy)_3^{2+}$, which was generated by Eq. 2. The return wave due to reduction of $Ru(bpy)_3^{3+}$ in the absence of DNA is not observed when DNA is present because any $Ru(bpy)_3^{3+}$ generated by the electrode is reduced by DNA and does not accumulate at the electrode. These observations were actually made first with $Re(O)_2(py)_4^{2+/+}$

[33], which exhibits a similar redox potential to $Ru(bpy)_3^{3+/2+}$ [34], but we later showed that $Ru(bpy)_3^{2+}$ was a more stable catalyst [17].

Numerous lines of evidence support the assignment of guanine as the electron donor. First, catalytic currents were obtained with genomic DNA and poly(GC)·poly(GC) but were not observed with poly(A)·poly(T) [17, 33]. Second, the electrode was poised at potentials sufficient to generate $Ru(bpy)_3^{3+}$ in the presence of radiolabeled oligonucleotides; when these oligonucleotides were treated with base and analyzed by high-resolution electrophoresis, selective cleavage at guanine was apparent [12, 17, 33, 35]. When we first observed this reaction, there were a few studies suggesting that one-electron oxidation of guanine produced piperidine-labile lesions [36, 37], and these lesions have been studied extensively in the intervening years [6, 8, 38].

The observation of the electrocatalysis in Fig. 1 suggests that the $Ru(bpy)_3^{3+/2+}$ and $guanine^{+/0}$ couples have similar redox potentials. Based on the kinetics of oxidation by a series of substituted $Ru(bpy)_3^{2+}$ complexes, we predicted that the redox potential of guanine was 1.1 V (all potentials versus Ag/AgCl) [17]. Later, equilibrium titrations performed by Steenken using known one-electron oxidants showed that the potential was 1.07 V at pH 7 [39], which also implied that the guanine deprotonates in our reaction. The issue of guanine deprotonation will be discussed in depth below.

As stated in the introduction, one of the attractions of electrochemistry as a mechanistic technique is the prospect of performing the analysis on small quantities of material [40]. Thus, we sought to develop a kinetic model whereby the catalytic currents could be simulated using the DigiSim software [9] to give rate constants for the guanine–Ru(III) electron transfer. The simplest case to analyze is that at high salt concentration, where binding of the $Ru(bpy)_3^{2+}$ to the DNA can be neglected [17, 18]. In this case, the kinetic model is simply that in Eqs. 1 and 2. In some cases, a second oxidation of $guanine_{ox}$ must be added to the model to generate acceptable fits [18]. Although the precise mechanism of guanine oxidation following one-electron transfer is still under investigation, it is generally accepted that this reaction would involve an overall oxidation of at least two electrons [41–43], so it is likely that any one-electron products would be susceptible to further reaction with Ru(III). The rate constant for the oxidation of guanine in genomic DNA was determined by simulation of the cyclic voltammograms using DigiSim to be 9,000 M^{-1} s^{-1}, and a similar value was obtained by analyzing cyclic square-wave voltammograms using the COOL algorithm [17]. The reaction was also studied using stopped-flow spectrophotometry with an authentic sample of $Ru(bpy)_3^{3+}$, and a similar value was again obtained [17].

3
Influence of DNA Binding

The initial kinetic simulations of the electrocatalytic cyclic voltammetry were performed for high salt conditions where the binding of $Ru(bpy)_3^{2+}$ to DNA could be neglected [17, 18, 33]. Parallel studies on the electrochemistry of the isostructural $Os(bpy)_3^{2+}$ $Os(bpy)_3^{2+}$ complex in the presence of DNA at low salt concentration show a number of effects of DNA binding on the electrochemical signals [44, 45]; these effects were initially noted by Carter and Bard in studies of $Co(phen)_3^{2+}$ $Co(phen)_3^{2+}$ [46, 47]. First, the DNA binding is tighter for the 3+ form than for the 2+ form due to the higher charge. Thus, the redox potential of the Os(III/II) couple is different for the bound and free forms [44]. Further, the diffusion coefficient of the bound form is the same as that for DNA, which is 2×10^{-7} cm²/s for genomic DNA compared with 7.3×10^{-6} cm²/s for free $Os(bpy)_3^{2+}$ [44, 45]. Thus, the majority of the electrochemistry occurs through the free $Os(bpy)_3^{3+/2+}$. A general square scheme showing the important bound and free states is shown in Fig. 2.

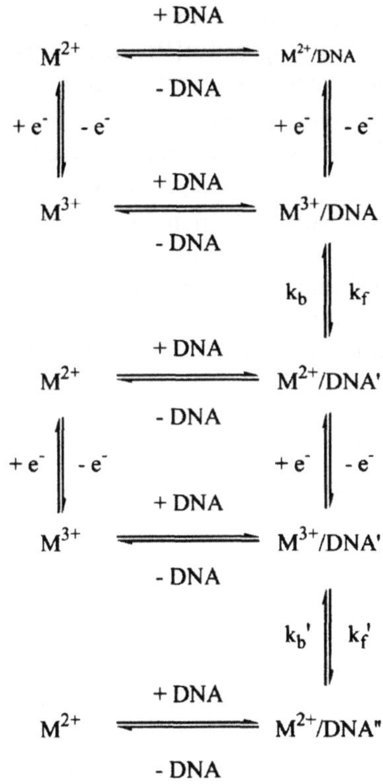

Fig. 2 Square schemes showing binding and oxidation by $M(bpy)^{3+/2+}$, where M is Os or Ru. Reprinted with permission from [19]. Copyright (1999) American Chemical Society

The effects of DNA binding on the catalytic oxidation are readily modeled using DigiSim, which allows input of multiple square schemes and different diffusion coefficients for the different couples [9]. Because the diffusion coefficients and binding constants are all known from the isostructural osmium case [44, 45], the catalytic oxidation with ruthenium can be modeled where the electron-transfer rates from guanine to bound Ru(III) are the only adjustable parameters [18, 19]. An additional effect that arises in the catalytic case is that the majority of the electrode oxidation occurs for the free $Ru(bpy)_3^{2+}$ complex, while DNA oxidation goes through the bound $Ru(bpy)_3^{3+}$. Effective second-order rate constants at low salt concentrations approach 10^6 M^{-1} s^{-1} [19].

Under low salt conditions, not only does the binding of the metal complex to the DNA need to be considered, but also the probability that once bound, an activated $Ru(bpy)_3^{3+}$ could abstract an electron from guanine [19]. For typical sequences that are 20–30% guanine, binding of the metal complex to the DNA would generally occur in relatively close proximity to a guanine such that electron transfer could be realized. However, to test this notion and the general robustness of our analysis, we determined the catalytic current for a 15-mer oligonucleotide duplex (**G15**, all sequences given in Table 1) that had only a single guanine at the center of one strand and no guanines on the complementary strand (Fig. 3). For this sequence, it can be readily envisioned that some binding of Ru(III) might occur close to the central guanine, which would result in electron transfer, while binding of Ru(III) at the distal end of the sequence might be too far from the central guanine for electron transfer to occur.

The simulation of the electrochemistry of $Ru(bpy)_3^{2+}$ in the presence of **G15** was performed with a model where each nucleotide was considered as a possible "active site". If the electrogenerated $Ru(bpy)_3^{3+}$ was bound at an active site nucleotide, then electron transfer would occur. Thus, for a 15-mer duplex, there are between one active site, where only binding at the guanine nucleotide would produce electron transfer, and 30 active sites, where binding at every nucleotide would produce electron transfer. As shown in Fig. 3, the models using one and 30 active sites under- and overestimate the cata-

Table 1 Oligonucleotide sequences

Name	Sequence (5′-3′)
G15	AAATATAGTATAAAA
GG16	AAATATAGGTATAAAA
GGG17	AAATATAGGGTATAAAA
GxG18	AAATATAGTAGTATAAAA
GxGxG21	AAATATAGTAGTAGTATAAAA
6G24	AAAAGTAGTAGTAGTAGTAGTAAA
3GG24	AAATATAGGTAGGTAGGTATAAAA
3GZ24	AAATATAGZTAGZTAGZTATAAAA
3ZG24	AAATATAZGTAZGTAZGTATAAAA
GQ2	AGGGTTAGGGTTAGGGTTAGGG
8G15	AAATATA8GTATAAAA

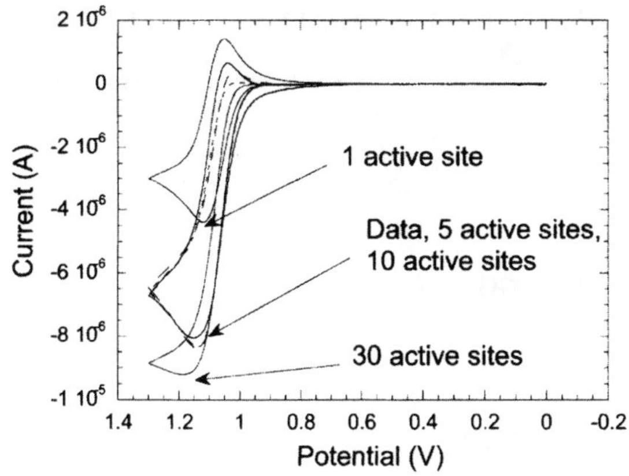

Fig. 3 Cyclic voltammograms of $Ru(bpy)_3^{2+}$ with **G15** hybridized to its complement (*dashed*). *Solid lines* show simulated currents obtained when productive binding of $Ru(bpy)^{2+}{}_3$ occurs at either 1, 5, 10, or 30 nucleotides ("active sites") in the duplex. Reprinted with permission from [19]. Copyright (1999) American Chemical Society

lytic current, respectively. The data could be simulated with models having either five or ten nucleotides, suggesting that electrons are transferred to a $Ru(bpy)_3^{3+}$ that is bound between 2.5 and 5 base pairs from the guanine donor. Increasing the number of guanines in the sequence increased the number of active sites in a predictable manner [19].

4 Effects of Sequence and Secondary Structure

4.1 Guanine Multiplets

An important observation made by a number of groups was that when two adjacent guanines were present in a DNA sequence, the 5′-guanine was preferentially oxidized [6, 48, 49]. This was typically observed for radiolabeled oligonucleotides reacted with one-electron oxidants and then piperidine-treated to reveal the sites of oxidation [6, 7, 38, 49, 50]. Upon visualization with high-resolution gel electrophoresis, more intense bands are observed at the 5′-guanine of the GG doublet. We sought to measure the relative reaction rates for guanine multiplets compared to that for guanine by cyclic voltammetry (CV) of oligonucleotides in the sequences 5′-AGT (**G15**), 5′-AGGT (**GG16**), and 5′-AGGGT (**GGG17**) [12]. The lowest reactivity on sequencing gels had been observed for a guanine with a T on the 3′-side [48], so we chose AGT as the sequence for an "isolated guanine". The voltammograms of $Ru(bpy)_3^{2+}$ in the presence of the sequences **G15**, **GG16**, and **GGG17** (Fig. 4a) were indicative of very rapid electron transfer for **GG16** and **GGG17** compared to **G15**.

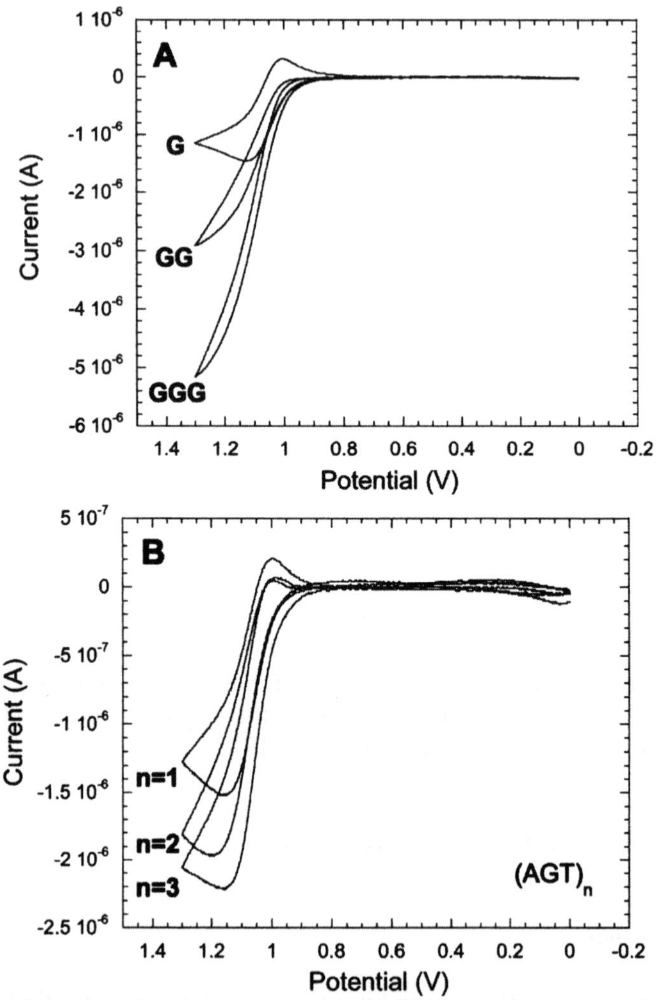

Fig. 4a,b Cyclic voltammograms of $Ru(bpy)_3^{2+}$ in the presence of DNA at pH 7. **a** Added sequences are the duplex forms of **G15** (G), **GG16** (GG), and **GGG17** (GGG). **b** Added sequences are the duplex forms of **G15** (*n*=1), **GxG18** (*n*=2), and **GxGxG21** (*n*=3). Reprinted with permission from [12]. Copyright (2000) American Chemical Society

Because the concentration of DNA strands was the same in each CV in Fig. 4a, the **GG16** and **GGG17** sequences contained twice and three times the concentration of guanine as **G15**, respectively. Since an increase in the absolute concentration of guanine also increases the catalytic current, we wanted to show that the faster rates for the guanine multiplets were not due simply to the increase in guanine concentration. We therefore examined the catalytic currents from oligonucleotides containing two and three isolated guanines, **GxG18** and **GxGxG21**, which contain 5′-AGTAGT and 5′-AGTAGTAGT sequences, respectively (Fig. 4b) [12]. Clearly, the increase in catalytic current is much greater for the guanine multiplet sequences than for the isolated guanine sequences with the same total number of guanines, so the enhanced reactivity in Fig. 4a is not due to the simple increase in guanine concentration that results from adding guanines to the sequence.

The apparent rate constant for oxidation of guanine could be obtained for simulations of the mediated voltammetry on the sequences **6G24** and **3GG24**, which both contain 24 total nucleotides and six guanines [12]. For the **6G24** oligonucleotide, all of the guanine bases were in an "isolated guanine" (AGT) sequence. For the **3GG24** oligonucleotide, however, there were two types of guanine present: 5′-A<u>G</u>G and 5′-G<u>G</u>T. The overall oxidation rate for **3GG24** was therefore the sum of the oxidation rates for the two types of guanine. An assumption that the rate constant for the 3′-G in the guanine doublet was

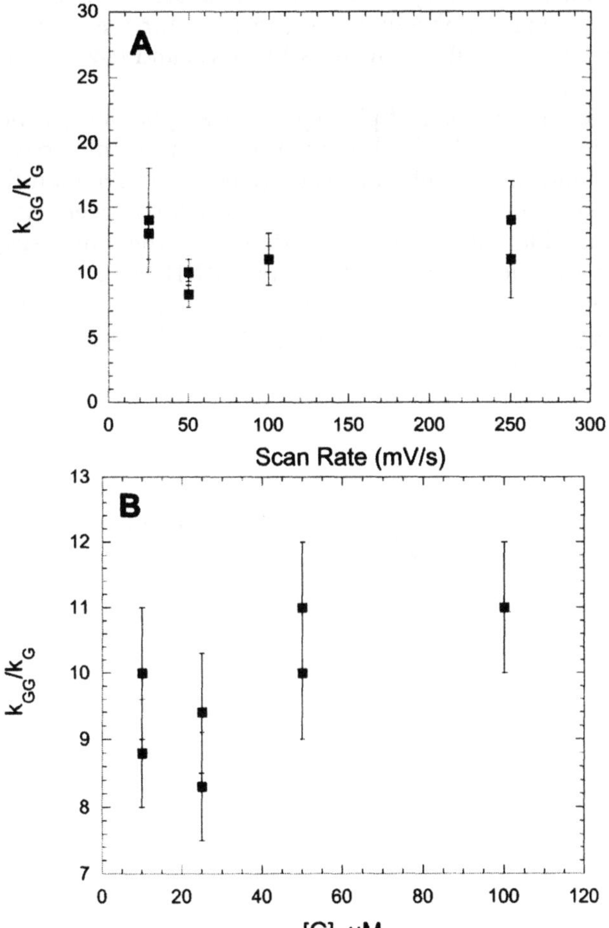

Fig. 5a,b Ratios of the rate constant for the 5′-G of the GG doublet to the 3′-G (k_{GG}/k_G) **a** at constant DNA concentration and **b** at constant scan rate. Reprinted with permission from [12]. Copyright (2000) American Chemical Society

identical to the rate constant for the isolated guanine (in **6G24**) is consistent with the observation of similar reactivities of the 3′-G in GG doublets on sequencing gels [51]. Thus, the observed rate constant for **3GG24** is:

$$k_{\mathrm{obs,3GG24}} = \frac{3k_{\mathrm{GG}} + 3k_{\mathrm{G}}}{6} \tag{3}$$

where k_{GG} is the rate constant for the 5′-G of the GG doublets and k_{G} is the rate constant for the 3′-G. This analysis gives a ratio for the 5′-G compared to isolated guanine, $k_{\mathrm{GG}}/k_{\mathrm{G}}$, of 12±2 across a broad range of scan rates (25 to 250 mV/s) and DNA concentrations (80 to 800 μM, Fig. 5). This ratio was generally lower than values obtained from band intensities on sequencing gels [52, 53], implying that the follow-up chemistry that leads from an oxidation event to strand scission exhibits a sequence selectivity that is the opposite of that for the primary oxidation. This observation shows the importance of quantitating the absolute rate constants for the primary oxidation. Later quantitation of similar sequences by Lewis and coworkers was consistent with the results in Fig. 5 [7].

We also studied the effect of placing a 7-deazaguanine (Z) on the 3′-side in doublets with guanine [12]. Although the 7-deazaguanine base is an excellent electron donor and exhibits a redox potential lower than that of native guanine [54, 55], 7-deazaguanine lacks N7, which has been hypothesized to provide the doublet stacking effect [56]. We therefore suspected that 7-deazaguanine would donate electrons to Ru(III) but that a 5′-GZ doublet might not show the same stacking effect as a 5′-GG doublet. The voltammograms of $Ru(bpy)_3^{2+}$ with **6G24**, **3GG24**, **3ZG24**, and **3GZ24** (Fig. 6) showed a

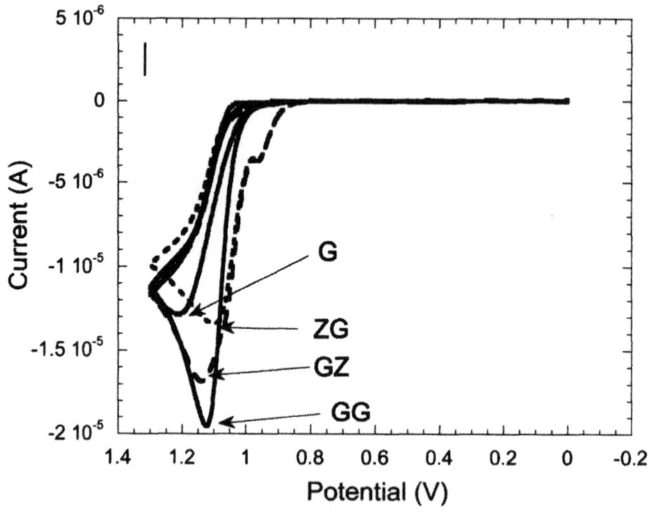

Fig. 6 Cyclic voltammograms of $Ru(bpy)_3^{2+}$ in the presence of **6G24** (G), **3ZG24** (ZG), **3GZ24** (GZ), and **3GG24** (GG). Reprinted with permission from [12]. Copyright (2000) American Chemical Society

similar amount of catalytic current for **6G24** and **6ZG24**, which was expected since both oligonucleotides contain six oxidizable bases. However, placing a guanine on the 5′-side of a 7-deazaguanine enhances the electron-transfer reactivity somewhat, but not to the same extent as native guanine. Thus, the stacking effect is not simply a measure of the overall electron density of the 3′-base, and the placement of the individual N7 atom is important.

4.2
Single-Base Mismatches

The studies on G multiplets clearly show that the sequence context of the oxidized guanine plays an important role in determining its oxidative reactivity [12]. We suspected that the secondary structure would also play an important role, both in determining the oxidation potential of the oxidized base and in dictating the accessibility of the base to solution-bound $Ru(bpy)_3^{3+}$. The rate constant for **G15** was determined both as single strand and hybridized to its perfect complement [17]. Hybridization to the perfect complement decreased the measured rate constant by a factor of 200, and the majority of this effect was due to protection of the oxidized guanine by the double helix, resulting in a larger electron-transfer distance. Placement of guanine in mismatches by hybridization to complements where the base opposite the G in **G15** was varied gave intermediate rate constants in the order GC<GT<GG<GA. A similarity of the rate constants for GT and GC was assigned to a lower distortion of the helix for a pyrimidine substitution than for a purine substitution.

4.3
G Quartets

Multistranded DNA structures are increasingly common and were first observed in early crystallography studies of nucleic acids [57, 58]. Guanine-rich sequences occur in telomeres at the ends of linear chromosomes and form G quartets where multiple guanines are organized around a central cation in a four-stranded structure (Fig. 7) [59]. As described above [12], DNA duplexes containing adjacent guanines show selective oxidation at the 5′-guanine of guanine multiplets [51], suggesting that guanine multiplets are traps for oxidizing equivalents in the DNA double helix [6, 52, 53, 60–63]. We were interested in whether G quartets formed from guanine triplets could be traps for oxidizing equivalents, so we studied electron transfer in an oligonucleotide **GQ2** that contains a repeating GGG sequence and forms a G quartet at the appropriate monovalent cation concentration [59, 64].

As shown in Fig. 8, oxidation by $Ru(bpy)_3^{3+}$ produces a current enhancement for the G quartet form of **GQ2** which is higher than for a duplex (**GD1**) containing the same number of guanines [35, 65]. The average rate constants *per guanine* are $(1.9\pm0.4)\times10^4$ M^{-1} s^{-1} for **GD1** (hybridized to its complement) and $(3.7\pm1.2)\times10^4$ M^{-1} s^{-1} for **GQ2** in its G quartet form, a much

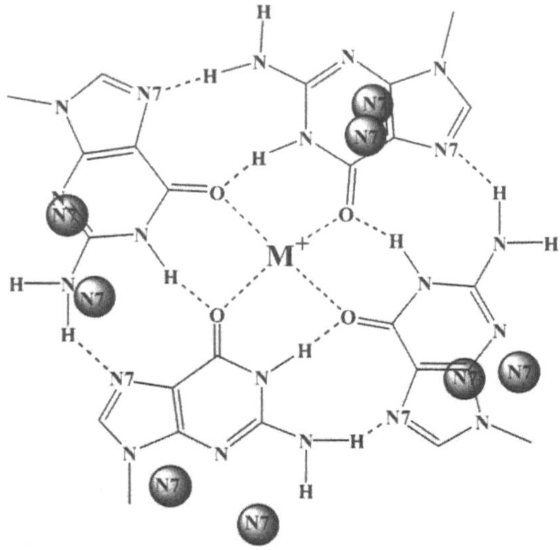

Fig. 7 View down the long axis of the top G quartet taken from coordinates given in [64]. The N7 atoms of the two lower quartets are shown as *gray spheres*. Reprinted with permission from [35]. Copyright (2000) American Chemical Society

smaller ratio than that observed for duplex GG sequences, which show an increase of a factor of 12 for the 5′-G [12]. To determine whether this increase was due to higher exposure of the guanines in the G quartet to $Ru(bpy)_3^{3+}$ or to higher reactivity as a result of the relative positions of the

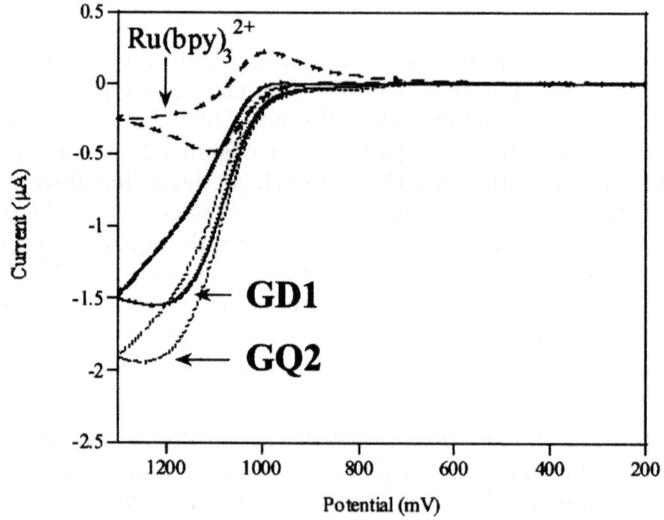

Fig. 8 Cyclic voltammograms of $Ru(bpy)_3^{2+}$ alone and with **GD1** and **GQ2**. Reprinted with permission from [35]. Copyright (2000) American Chemical Society

guanines in the G quartets, we performed high-resolution gel electrophoresis to determine the extent of oxidation at each guanine in sequence **GQ3**, which is identical to **GQ2** except for the addition of a single-stranded TGT triplet to the 3′ end (Fig. 9). The oxidation was performed by the "flash-quench" technique [66], where $Ru(bpy)_3^{2+}$ is photolyzed in the presence of $Fe(CN)^{3-}{}_6$ to generate $Ru(bpy)_3^{3+}$, the same oxidant used in our electrochemical measurements. The G quartet form gave a selectivity pattern of 5′>3′>central in contrast to the normal pattern for the duplex form where the 5′-guanines are enhanced in reactivity compared to the 3′-guanine. As shown in Fig. 9, when **GQ3** was hybridized to its complement, the 5′-G enhancement returned. Thus, there was no evidence of the increase in 5′-guanine reactivity as seen for duplex DNA, consistent with the electrochemical results.

The guanines in G quartets are slightly more accessible to the Ru(III) oxidant than in B-form duplexes, but stacking of adjacent guanines in the G quartet does not provide an increase in oxidative reactivity. The average solvent-accessible surface areas for guanines in duplex, G quartet, and single strands (148, 164, and 253 $Å^2$, respectively) follow the same trend as our measured rate constants (1.9×10^4 M^{-1} s^{-1}, 3.7×10^4 M^{-1} s^{-1}, and 2×10^5 M^{-1} s^{-1}, respectively). This result supports the hypothesis that the increase in electron density of the 5′-guanine in B-DNA results from the position of the N7 of the 3′-guanine relative to the π-system of the 5′-guanine [56], which is also consistent with the 7-deazaguanine chemistry shown in Fig. 6. As shown in Fig. 7, there is no similar alignment of the N7 atoms in the G quartet structure. This is consistent with the observation of Schuster et al. that there is less selectivity for the 5′-G in a guanine doublet in A-form DNA:RNA hybrids [62]. Because adjacent guanines in G quartets are not effective hole traps, formation of the G quartet structure might protect guanine multiplets from oxidation in vivo [63, 67].

5
Effect of Deprotonation

The coupling of a proton transfer to an electron transfer has been recognized as a critical issue in redox kinetics and thermodynamics [68–74]. An electron-transfer reaction that produces an oxidized product that is acidic leads to deprotonation of the oxidized species, lowering the overall energy of the reaction according to the Nernst equation. When the proton-transfer and electron-transfer events are separated, the first step is often highly unfavorable, involving either initial electron transfer to form a highly acidic oxidized species or initial proton transfer to form a strongly reducing deprotonated species. Therefore, these reactions often occur via a concerted proton-coupled electron transfer (PCET, Fig. 10) [69, 71, 74]. The PCET route influences the kinetic barrier to the overall process and often leads to large isotope effects [72, 73, 75, 76]. The factors that control the kinetic barriers to the PCET process have been assessed both experimentally and theoretically by many research groups [68–75].

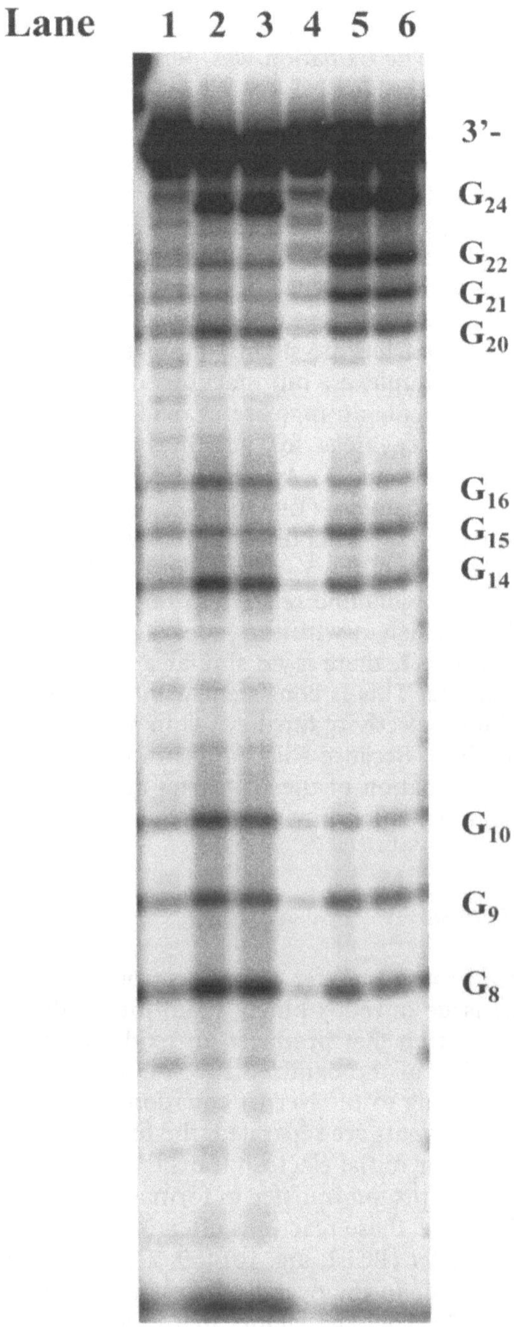
Lane 1 2 3 4 5 6
3'-
G24
G22
G21
G20
G16
G15
G14
G10
G9
G8

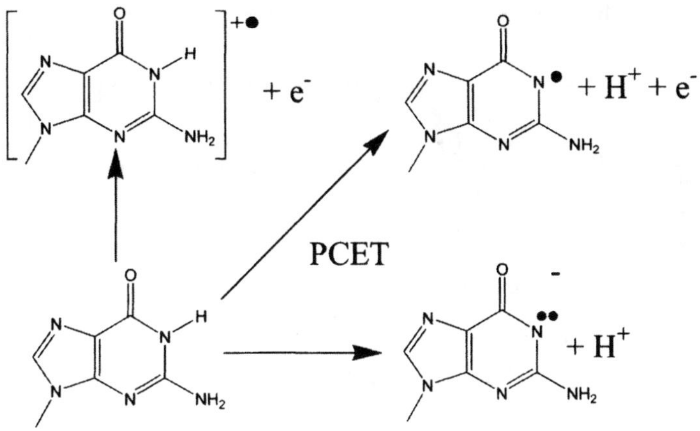

Fig. 10 Reaction scheme for PCET. Figure taken from reference [78]

We have performed a number of studies suggesting that the electron-transfer reaction of $Ru(bpy)_3^{3+}$ with guanine is a proton-coupled reaction [77, 78]. The guanine–Ru(III) reaction proceeds with an isotope effect of about 2, seen most strikingly by comparing the catalytic currents obtained in H_2O and D_2O (Fig. 11). This result is consistent with cleavage studies by Giese and Wessely [8] and photochemical measurements by Shafirovich et al. [79–81]. The isotope effect falls in the range of 1.4–2 for guanine in a free mononucleotide or hydrogen-bonded in duplex DNA, implying that the proton coupling can be realized from within the fully formed double helix. Recent experiments show that the rate constant exhibits a linear dependence on the quantity of isotope present (Fig. 12) [78], implying that there is a single proton associated with the reaction [82]. As expected, linear plots are observed for both duplex DNA and free mononucleotide, showing that the hybridization state does not affect the importance of the proton in the reaction.

Fig. 9 Phosphorimage of a denaturing polyacrylamide gel showing the results of photolytic cleavage of **GQ3** with $Ru(bpy)_3^{2+}$ and $Fe(CN)_6^{3-}$ in potassium phosphate and 50 mM added KCl. Lane 1: G quartet **GQ3**, no $Ru(bpy)_3^{2+}$ or $Fe(CN)_6^{3-}$. Lane 2: G quartet **GQ3**, 50 μM $Ru(bpy)_3^{2+}$, 500 μM $Fe(CN)_6^{3-}$. Lane 3: G quartet **GQ3**, 100 μM $Ru(bpy)_3^{2+}$, 1 mM $Fe(CN)_6^{3-}$. Lane 4: duplex **GQ3**, no $Ru(bpy)_3^{2+}$, no $Fe(CN)_6^{3-}$. Lane 5: duplex **GQ3**, 50 μM $Ru(bpy)_3^{2+}$, 500 μM $Fe(CN)_6^{3-}$. Lane 6: duplex **GQ3**, 100 μM $Ru(bpy)_3^{2+}$, 1 mM $Fe(CN)_6^{3-}$. Enhanced reactivity at G_{22} in lanes 5 and 6 is due to partial fraying of the duplex; reaction of **GQ3** hybridized to its full-length complement (i.e., where the TGT overhang was hybridized) gave the same pattern at G_{20}–G_{22} as the other two triplets. Reprinted with permission from [35]. Copyright (2000) American Chemical Society

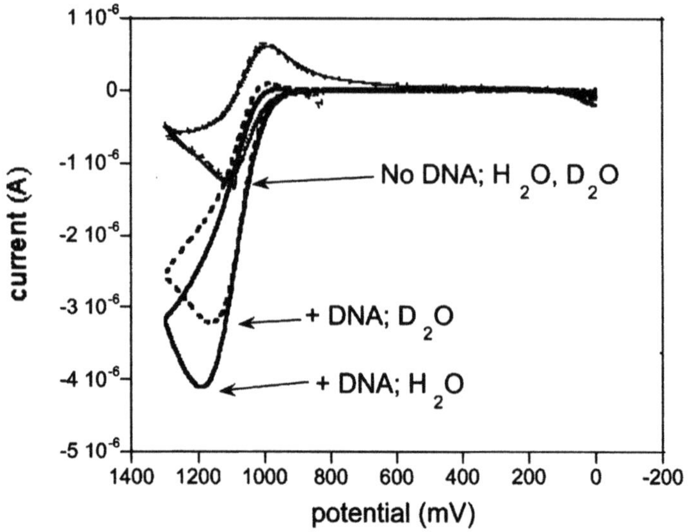

Fig. 11 Cyclic voltammograms of $Ru(bpy)_3^{2+}$ with herring testes DNA in H_2O and D_2O. Reprinted with permission from [77]. Copyright (2001) American Chemical Society

We have also shown that the dependence of the oxidation rate on driving force in H_2O gives a plot of $RT \ln k$ versus $E_{1/2}$ of the metal oxidant with a slope of 0.8±0.1 [77], significantly higher than the theoretically predicted value of 0.5 for a simple electron-transfer reaction [83]. The elevated slope likely results from the involvement of the proton in the reaction. Like the isotope effect, this high slope was observed for guanine mononucleotides, single-stranded oligonucleotides containing guanine, duplex oligonucleotides containing guanine, and genomic DNA [77]. Thus, the concerted PCET pathway is followed for guanine that is highly accessible to the solvent and for guanine that is enclosed in the double helix and base-paired to cytosine. These studies address an important general question in biological electron transfer involving redox sites that are buried in biomolecules but that require proton transfer to undergo oxidation. This question pertains to oxidation of buried organic sites in proteins [84] in addition to DNA oxidation. Both the high slope and the isotope effect imply that the guanine is able to transfer its proton to the solvent from within the double helix concomitantly with the electron transfer. Related studies are those showing that tyrosine and tryptophan residues can readily deprotonate from within the hydrophobic cores of redox proteins [85].

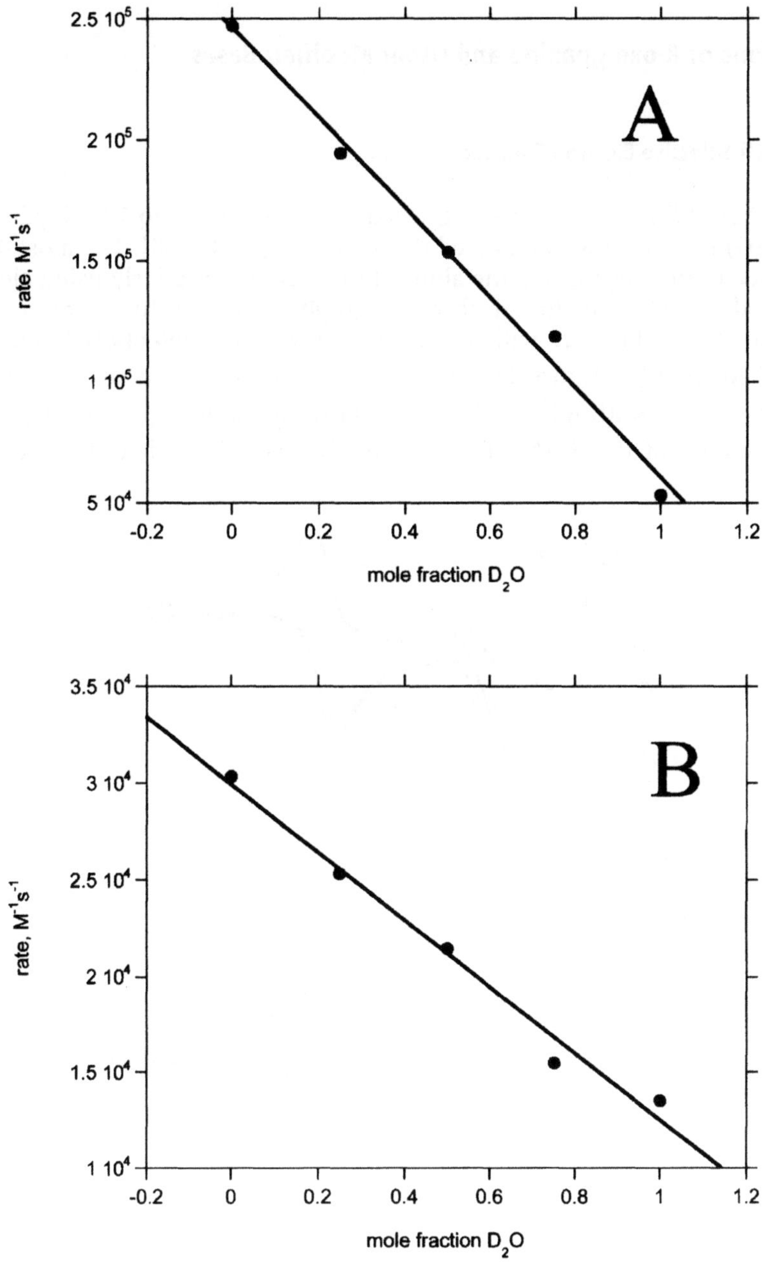

Fig. 12a,b Electron-transfer rate constants obtained by fitting cyclic voltammograms of **a** dGMP and 50 μM $Ru(bpy)_3^{2+}$ and **b** 1 mM ht DNA and 50 μM $Ru(bpy)_3^{2+}$ in 800 mM NaCl with varying percentages of D_2O. Reprinted with permission from [78]. Copyright (2003) American Chemical Society

6 Reactions of 8-oxo-guanine and Other Modified Bases

6.1 Mismatch-Selective Electron Transfer

A number of lines of evidence suggested early on that 8-oxo-7,8-dihydroguanine (8G) is more readily oxidized than guanine [42, 86–88]. The most direct evidence in this regard was the ability to detect 8G selectively using electrochemical detectors in liquid chromatography [89]. We therefore suspected that 8G would react with metal complexes with potentials lower than that of $Ru(bpy)_3^{2+}$. Figure 13 shows cyclic voltammograms of a mixture of $Os(bpy)_3^{2+}$ and $Ru(bpy)_3^{2+}$. When only native guanine is present (Fig. 13a), enhancement of the $Ru(bpy)_3^{2+}$ wave is observed, but when 8G is present

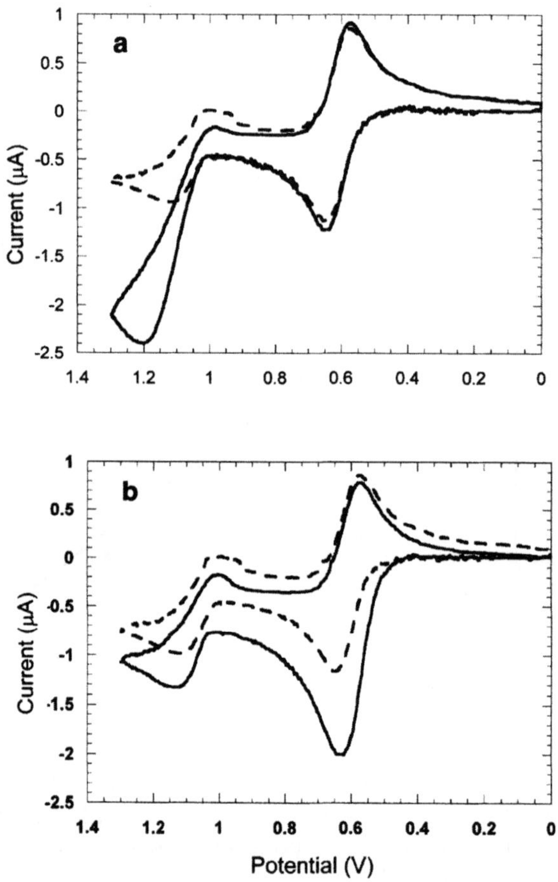

Fig. 13a,b Cyclic voltammograms of mixtures of $Os(bpy)_3^{2+}$ and $Ru(bpy)_3^{2+}$ alone (*dashed*) and with **a** G15 and **b** 8G15 (*solid*). Figure modified from reference [101]

(Fig. 13b), enhancement of the $Os(bpy)_3^{3+/2+}$ wave is observed. As with native guanine and $Ru(bpy)_3^{2+}$, the current enhancement is significantly larger for a single-stranded oligomer than when hybridized to its complement, and the duplexes with mismatches at the 8G give intermediate degrees of current enhancement. Simulation of the voltammograms shows that the rate of oxidation of 8G hybridized to a C by Os(III) is approximately an order of magnitude faster than for oxidation of guanine by Ru(III). The selectivity of Os(III) for 8G is striking: even when an oligonucleotide contained the electron-rich 5′-GG doublet and a 5′-GGG triplet, there was a large enhancement in the Ru(III/II) wave but no detectable enhancement in the Os(III/II) couple. Steenken et al. later published an equilibrium titration showing that the potential of 8G was 0.50 V, consistent with our observations [90].

6.2 Detection of a Physiologically Important Deletion

An attraction of the mismatch selectivity observed for oxidation of guanine by Ru(III) is the potential to use such a reaction in the detection of single-base mismatches in clinical samples [17]; however, in the guanine–Ru(III) reaction, other guanines in the sequence obscure the effect of the mismatch on the target site. We therefore reasoned that the 8G–Os(III) reaction might allow for detection of a mismatch at 8G in the presence of native guanines. The ability to detect selectively the secondary structure at a specific 8G was tested in a physiologically relevant sequence using an oligomer that is complementary to a site of common genetic mutation for cystic fibrosis (CF) [91]. Deletion of a phenylalanine codon (TTT) from this gene is responsible for 70% of all cases of CF [91]. We designed a probe oligonucleotide with the sequence 5′-ATAGGAAACACC-A8GA-GATGATATTTTC where the wild-type complement contains a 5′-TTT triplet opposite the A8GA on the probe. The A8GA sequence gives a relatively small enhancement when hybridized to the full-length (+TTT) complement, while the physiologically relevant −TTT mutant places the 8G in a single-stranded A8GA bulge, which gives a significantly higher current enhancement. This experiment provides a strategy for detecting point mutations by designing probes modified with 8G that can be selectively oxidized by $Os(bpy)_3^{3+}$ in the presence of native guanines.

6.3 Computational Studies

The insight provided by studying 8-oxo-guanine, and the ability to substitute DNA with a nucleobase that could be selectively oxidized by a low-potential complex, prompted us to search for other minimally substituted, redox-active nucleobases [92]. We therefore developed a library of nucleobases that were investigated using density functional theory (DFT) [93, 94] calculations self-consistently coupled to the conductorlike solvation model (COSMO) [95, 96]. The case of oxidation of nucleobases, particularly guanine, had been examined previously by similar calculations [48, 56]; however, we sought to model not just the simple one-electron oxidation but also the effect of deprotonation of the one-electron oxidation product [39, 81, 90, 97], discussed above. In fact, calculation of the simple one-electron potentials did not produce the appropriate trend in observed reactivities [92]. Our goal was therefore to predict the standard redox potentials of the proton-coupled redox processes for the overall reaction:

$$B \xrightarrow{-e^-} B^+ \rightarrow B(-H) + H^+ \qquad (4)$$

Only the first deprotonation product of the cationic nucleobase radical needed to be considered to obtain the appropriate trends in potential, which were verified by testing whether electrocatalytic oxidation was observed with metal complexes of appropriate potential.

Extending the electrocatalytic approach to pyrimidine bases was an important goal that would greatly expand the number of substitutions possible for developing highly redox-active biomolecules. The computational library suggested that the modified pyrimidines 5-aminocytosine, 5-aminouracil, and 6-aminocytosine would likely be oxidizable in our system. The calculations suggested that 5-aminocytosine would exhibit a potential 0.58 eV lower than that of guanine and even lower than that of 8-oxo-guanine. A similar shift was predicted for 5-aminouracil. The large, unexpected shift observed upon addition of the amino group at the 5-position suggested that these compounds would react with $Os(bpy)_3^{2+}$, which was confirmed experimentally. Substitution of an amino group at the 6-position of cytosine also produced a significant shift for 6-aminocytosine, which accordingly gave current enhancement with $Ru(bpy)_3^{2+}$ but not $Os(bpy)_3^{2+}$ [92].

7 Perspectives on Long-Range Electron Transfer

A number of observations discussed here are germane to the questions surrounding electron transfer in nucleic acids over long distances. In fact, all of the observations made in our system are best interpreted in terms of relatively short-range reactions. Before discussing these points in detail, two important caveats must be discussed. The first is that in our case, the oxidant (generally $Ru(bpy)_3^{2+}$) is not intercalated; many of the systems discussed in this volume involve intercalated donors and acceptors. Thus, if intercalation is a requirement for strong electronic coupling between donors and acceptors across DNA, these conditions will not obtain in our system. Second, the timescale of our reactions is reasonably long. The electron transfers occur in milliseconds, which are readily monitored by electrochemistry and stopped-flow spectrophotometry. This timescale results partly from the relatively low exergonicity of our reactions, which occurs generally near $\Delta G^0=0$. Again, if reactions that are much faster are required (and hence at much higher driving forces) to observe long-range effects, then these conditions do not obtain for our system. Many of the systems discussed in this volume are those where the timescale for electron transfer is much faster.

With these caveats in mind, a number of points support a relatively short distance for guanine–$Ru(bpy)_3^{3+}$ electron transfer in our system. First, as shown in Fig. 3, when the guanines are scarce in the sequence, the binding of $Ru(bpy)_3^{3+}$ at sites on the helix to which electron transfer is not efficient must be considered [19]. Thus, the oxidant can bind to the duplex at sites too far from a guanine for electron transfer. Quantitative analysis of these data suggests that electron transfer occurs within distances that are consistent with those observed in studies of covalently bound donors and acceptors and therefore with values of the tunneling parameter β consistent with those observed in other systems [38, 50, 98–100].

The second point supporting a short-range reaction is the effect of secondary structure [17, 101, 102]. As discussed above, the reaction efficiency is subject to variations in the structure of guanine induced by single-base mismatches [17, 101, 102], hairpins and bulges [101], and G quartets [35]. These experiments show that when the solvent accessibility at guanine is increased by distortions in the nucleic acid structure near the oxidized guanine, the rate of electron transfer increases. If the reaction were occurring over long distances, it is likely that the local geometry at guanine would not be important since the electron transfer could be occurring to an oxidant bound at a site remote from that of the oxidized guanine.

The third point of note regarding long-range electron transfer in this system is the clear importance of the proton transfer in the reaction [77, 78]. The isotope effects and effects of driving force point to the coupling of a deprotonation to the guanine oxidation. Giese has discussed how deprotonation can control the rates of electron movement in hole transfer through the helix [8]. Our observation that proton transfer occurs concomitantly with oxidation of guanine by $Ru(bpy)_3^{3+}$ is consistent with Giese's proposal. Similarly, perusal of the chapters in this volume shows that a unified picture of electron transfer in DNA is emerging. The consistency of our electrocatalytic studies involving freely diffusing metal complex oxidants with studies of electron transfer in attached donors and acceptors is indicative of the maturation of the field.

Acknowledgments I thank my numerous energetic coworkers who have contributed to this work and whose names are given in the cited references. Work in my laboratory on guanine oxidation has been supported by the NSF, Department of Defense, and Xanthon, Inc.

References

1. Beckman KB, Ames BN (1997) J Biol Chem 272:19633
2. Popovich N, Thorp HH (2002) Interface 11:30
3. Kuhr WG (2000) Nature Biotech 18:1042
4. Wang J, Kawde A-N, Sahlin E, Parrado C, Rivas G (2000) Electroanalysis 12:917
5. Palecek E (2002) Talanta 56:809
6. Hall DB, Holmlin RE, Barton JK (1996) Nature 382:731
7. Lewis FD, Liu X, Liu J, Hayes RT, Wasielewski MR (2000) J Am Chem Soc 122:12037
8. Giese B, Wessely S (2001) Chem Commun 2108

9. Rudolph M, Reddy DP, Feldberg SW (1994) Anal Chem 66:589a
10. Osteryoung J (1993) Acc Chem Res 26:77
11. Bard AJ, Faulkner LR (1980) Electrochemical methods. Wiley, New York
12. Sistare MF, Codden SJ, Heimlich G, Thorp HH (2000) J Am Chem Soc 122:4742
13. Palecek E (1996) Electroanalysis 8:7
14. Brazill SA, Kim PH, Kuhr WG (2001) Anal Chem 73:4882
15. Singhal P, Kuhr WG (1997) Anal Chem 69:4828
16. Brazill SA, Singhal P, Kuhr WG (2000) Anal Chem 72:5542
17. Johnston DH, Glasgow KC, Thorp HH (1995) J Am Chem Soc 117:8933
18. Johnston DH, Thorp HH (1996) J Phys Chem 100:13837
19. Sistare MF, Holmberg RC, Thorp HH (1999) J Phys Chem B 103:10718
20. Yang IV, Thorp HH (2001) Anal Chem 73:5316
21. Napier ME, Thorp HH (1997) Langmuir 13:6342
22. Armistead PM, Thorp HH (2001) Anal Chem 73:558
23. Armistead PM, Thorp HH (2000) Anal Chem 72:3764
24. Yang IV, Ropp PA, Thorp HH (2002) Anal Chem 74:347
25. Szalai VA, Jayawickamarajah J, Thorp HH (2002) J Phys Chem B 106:709
26. Yang IV, Thorp HH (2000) Inorg Chem 39:4969
27. Szalai VA, Thorp HH (2000) J Phys Chem B 104:6851
28. Armstrong NR, Lin AWC, Fujihira M, Kuwana T (1976) Anal Chem 48:741
29. Popovich ND, Wong SS, Yen BKH, Paine DC (2002) Anal Chem 74:3127
30. Grover N, Thorp HH (1991) J Am Chem Soc 113:7030
31. Willit JL, Bowden EF (1990) J Phys Chem 94:8241
32. Wang J, Pakecek E, Nielsen PE, Rivas G, Cai X, Shiraishi H, Dontha N, Luo D, Farias PAM (1996) J Am Chem Soc 118:7667
33. Johnston DH, Cheng C-C, Campbell KJ, Thorp HH (1994) Inorg Chem 33:6388
34. Thorp HH, Van Houten J, Gray HB (1989) Inorg Chem 28:889
35. Szalai VA, Thorp HH (2000) J Am Chem Soc 122:4524
36. Federova OS, Podust LM (1988) J Inorg Biochem 34:149
37. Kasai H, Yamgizumi Z, Berger M, Cadet J (1992) J Am Chem Soc 114:9692
38. Schuster GB (2000) Acc Chem Res 33:253
39. Steenken S, Jovanovic SV (1997) J Am Chem Soc 119:617
40. Wightman RM (1981) Anal Chem 53:1125A
41. Cadet J, Bellon S, Berger M, Bourdat AG, Douki T, Duarte V, Frelon S, Gasparutto D, Muller E, Ravanat JL, Sauvaigo S (2002) Biol Chem 383:933
42. Doddridge ZA, Cullis PM, Jones GDD, Malone ME (1998) J Am Chem Soc 120:10998
43. Burrows CJ, Muller JG (1998) Chem Rev 98:1109
44. Welch TW, Thorp HH (1996) J Phys Chem 100:13829
45. Welch TW, Corbett AH, Thorp HH (1995) J Phys Chem 99:11757
46. Carter MJ, Bard AJ (1987) J Am Chem Soc 109:7528
47. Carter MT, Rodriguez M, Bard AJ (1989) J Am Chem Soc 111:8901
48. Sugiyama H, Saito I (1996) J Am Chem Soc 118:7063
49. Saito I, Nakamura T, Nakatani K, Yoshioka Y, Yamaguchi K, Sugiyama H (1998) J Am Chem Soc 120:12686
50. Meggers E, Michel-Beyerle ME, Giese B (1998) J Am Chem Soc 120:12950
51. Saito I, Takayama M, Sugiyama H, Nakatani K, Tsuchida A, Yamamoto M (1995) J Am Chem Soc 117:6406
52. Hickerson RP, Prat F, Muller JG, Foote CS, Burrows CJ (1999) J Am Chem Soc 121:9423
53. Kan Y, Schuster GB (1999) J Am Chem Soc 121:10857
54. Wan C, Fiebig T, Kelley SO, Treadway CR, Barton JK, Zewail AH (1999) Proc Natl Acad Sci USA 96:6014
55. Kelley SO, Barton JK (1999) Science 283:375
56. Prat F, Houk KN, Foote CS (1998) J Am Chem Soc 120:845
57. Saenger W (1984) Principles of nucleic acid structure. Springer, Berlin Heidelberg New York

58. Gilbert DE, Feigon J (1999) Curr Opin Struct Biol 9:305
59. Williamson JR, Raghuraman MK, Cech TR (1989) Cell 59:871
60. Kan Y, Schuster GB (1999) J Am Chem Soc 121:11607
61. Ly D, Sanii L, Schuster GB (1999) J Am Chem Soc 121:9400
62. Sartor V, Henderson PT, Schuster GB (1999) J Am Chem Soc 121:11027
63. Nunez ME, Hall DB, Barton JK (1999) Chem Biol 6:85
64. Wang Y, Patel DJ (1993) Structure 1:263
65. Szalai VA, Singer MJ, Thorp HH (2002) J Am Chem Soc 124:1625
66. Stemp EDA, Arkin MR, Barton JK (1997) J Am Chem Soc 119:2921
67. Henderson PT, Jones D, Hampikian G, Kan Y, Schuster GB (1999) Proc Natl Acad Sci USA 96:8353
68. Roth JP, Lovell S, Mayer JM (2000) J Am Chem Soc 122:5486
69. Cukier RI, Nocera DG (1998) Annu Rev Phys Chem 49:337
70. Turro C, Zaleski M, Karabatsos YM, Nocera DG (1996) J Am Chem Soc 118:6060
71. Hammes-Schiffer S (2001) Acc Chem Res 34:273
72. Iordanova N, Decornez H, Hammes-Schiffer S (2001) J Am Chem Soc 123:3723
73. Iordanova N, Hammes-Schiffer S (2002) J Am Chem Soc 124:4848
74. Meyer TJ (1984) J Electrochem Soc 131:221C
75. Farrer BT, Thorp HH (1999) Inorg Chem 38:2497
76. Stultz LK, Binstead RA, Reynolds MS, Meyer TJ (1995) J Am Chem Soc 117:2520
77. Weatherly SC, Yang IV, Thorp HH (2001) J Am Chem Soc 123:1236
78. Weatherly SC, Yang IV, Armistead PM, Thorp HH (2003) J Phys Chem B 107:372
79. Shafirovich V, Dourandin A, Geacintov NE (2001) J Phys Chem B 105:8431
80. Shafirovich V, Dourandin A, Huang W, Luneva NP, Geacintov NE (1999) J Phys Chem B 103:10924
81. Shafirovich V, Dourandin A, Luneva NP, Geacintov NE (2000) J Phys Chem B 104:137
82. Binstead RA, Meyer TJ (1987) J Am Chem Soc 109:3287
83. Marcus RA, Sutin N (1985) Biochim Biophys Acta 811:265
84. Chan SI, Li PM (1990) Biochemistry 29:1
85. Di Bilio AJ, Crane BR, Wehbi WA, Kiser CN, Abu-Omar MM, Carlos RM, Richards JH, Winkler JR, Gray HB (2001) J Am Chem Soc 123:3181
86. Bernstein R, Prat F, Foote CS (1999) J Am Chem Soc 121:464
87. Muller JG, Duarte V, Hickerson RP, Burrows CJ (1998) Nucleic Acids Res 26:2247
88. Koizume S, Inoue H, Kamiya H, Ohtsuka E (1996) Chem Commun 265
89. Helbock HJ, Beckman KB, Shigenaga MK, Walter PB, Woodall AA, Yeo HC, Ames BN (1998) Proc Natl Acad Sci USA 95:288
90. Steenken S, Jovanovic SV, Bietti M, Bernhard K (2000) J Am Chem Soc 122:2373
91. Riordan JR, Rommens JM, Kerem B, Alon N, Rozmahel R, Grzelczak Z, Zielenski J, Lok S, Plavsic N, Chou J-L, Drumm ML, Iannuzzi MC, Collins FS, Tsui L-C (1989) Science 245:1066
92. Baik M-H, Silverman JS, Yang IV, Ropp PA, Szalai VA, Yang W, Thorp HH (2001) J Phys Chem B 105:6537
93. Parr RG, Yang W (1989) Density functional theory of atoms and molecules. Oxford University Press, New York
94. Ziegler T (1991) Chem Rev 91:651
95. Klamt A, Schuurmann G (1993) J Chem Soc Perkin Trans 2:799
96. Pye CC, Ziegler T (1999) Theor Chim Acta 101:396
97. Steenken S (1997) Biol Chem 378:1293
98. Meade TJ, Kayyem JF (1995) Angew Chem Int Ed Engl 34:352
99. Lewis FD, Wu T, Zhang Y, Letsinger RL, Greenfield SR, Wasielewski MR (1997) Science 277:673
100. Jortner J, Bixon M, Langenbacher T, Michel-Beyerle ME (1998) Proc Natl Acad Sci USA 95:12759
101. Ropp PA, Thorp HH (1999) Chem Biol 6:599
102. Yang IV, Thorp HH (2001) Inorg Chem 40:1690

Top Curr Chem (2004) 237:183–227
DOI 10.1007/b94477

Charge Transport in DNA-Based Devices

Danny Porath[1] · Gianaurelio Cuniberti[2] · Rosa Di Felice[3]

[1] Department of Physical Chemistry, Institute of Chemistry, The Hebrew University, 91904 Jerusalem, Israel
E-mail: porath@chem.ch.huji.ac.il

[2] Institute for Theoretical Physics, University of Regensburg, 93040 Regensburg, Germany
E-mail: g.cuniberti@physik.uni-regensburg.de

[3] INFM Center on nanoStructures and bioSystems at Surfaces (S3), Università di Modena e Reggio Emilia, Via Campi 213/A, 41100 Modena, Italy
E-mail: rosa@unimore.it

Abstract Charge migration along DNA molecules has attracted scientific interest for over half a century. Reports on possible high rates of charge transfer between donor and acceptor through the DNA, obtained in the last decade from solution chemistry experiments on large numbers of molecules, triggered a series of direct electrical transport measurements through DNA single molecules, bundles, and networks. These measurements are reviewed and presented here. From these experiments we conclude that electrical transport is feasible in short DNA molecules, in bundles and networks, but blocked in long single molecules that are attached to surfaces. The experimental background is complemented by an account of the theoretical/computational schemes that are applied to study the electronic and transport properties of DNA-based nanowires. Examples of selected applications are given, to show the capabilities and limits of current theoretical approaches to accurately describe the wires, interpret the transport measurements, and predict suitable strategies to enhance the conductivity of DNA nanostructures.

Keywords Molecular electronics · Biomolecular nanowires · Conductance · Bandstructure · Direct electrical transport

Abbreviations

Ade (A)	Adenine
Cyt (C)	Cytosine
Gua (G)	Guanine
Thy (T)	Thymine
1D	One-dimensional
AFM	Atomic force microscope
BLYP	Becke–Lee–Yang–Parr (GGA)
BZ	Brillouin zone
CNT	Carbon nanotube
DFT	Density functional theory
DOS	Density of states
EFM	Electrostatic force microscope
GGA	Generalized gradient approximation
HF	Hartree–Fock
HOMO	Highest occupied molecular orbital
LDA	Local density approximation
LEEPS	Low-energy electron point source
LUMO	Lowest unoccupied molecular orbital
MP2	Møller–Plesset 2nd order
NMR	Nuclear magnetic resonance
PBE	Perdew–Burke–Ernzerhof (GGA)
SEM	Scanning electron microscope
SFM	Scanning force microscope

STM	Scanning tunneling microscope
TB	Tight binding
TEM	Transmission electron microscope

1 Introduction

1.1 Devices go Molecular—the Emergence of Molecular Electronics

The progress of the electronic industry in the past few decades was based on the delivery of smaller and smaller devices and denser integrated circuits, which ensured the attainment of more and more powerful computers. However, such a fast growth is compromised by the intrinsic limitations of the conventional technology. Electronic circuits are currently fabricated with complementary-metal-oxide-semiconductor (CMOS) transistors. Higher transistor density on a single chip means faster circuit performance. The trend towards higher integration is restricted by the limitations of the current lithography technologies, by heat dissipation, and by capacitive coupling between different components. Moreover, the down-scaling of individual devices to the nanometer range collides with fundamental physical laws. In fact, in conventional silicon-based electronic devices the information is carried by mobile electrons within a band of allowed energies according to the semiconductor bandstructure. However, when the dimensions shrink to the nanometer scale, and bands turn into discrete energy levels, then quantum correlation effects induce localization.

In order to pursue the miniaturization of integrated circuits further [1], a novel technology, which would exploit the pure quantum mechanical effects that rule at the nanometer scale, is therefore demanded. The search for efficient molecular devices, that would be able to perform operations currently done by silicon transistors, is pursued within this framework. The basic idea of molecular electronics is to use individual molecules as wires, switches, rectifiers, and memories [2–6]. Another conceptual idea that is advanced by molecular electronics is the switch from a top-bottom approach, where the devices are extracted from a single large-scale building block, to a bottom-up approach in which the whole system is composed of small basic building blocks with recognition, structuring, and self-assembly properties. The great advantage of molecular electronics in the frame of the continued device miniaturization is the intrinsic nanoscale size of the molecular building blocks that are used in the bottom-up approach, as well as the fact that they may be synthesized in parallel in huge quantities and at low cost. Different candidates for molecular devices are currently the subject of highly interdisciplinary investigation efforts, including small organic polymers [6–11], large biomolecules [12–20], nanotubes, and fullerenes [21–24]. In the following, we focus on the exploration of DNA biomolecules as prospective candidates for molecular electronic devices.

For the scientists devoted to the investigation of charge mobility in DNA, a no less important motivation than the strong technological drive is that DNA molecules comprise an excellent model system for charge transport in one-dimensional polymers. This most well-known polymer enables an endless number of structural manipulations in which charge transport mechanisms like hopping and tunneling may be studied in a controlled way.

1.2 The Unique Advantages of DNA-Based Devices—Recognition and Structuring

Two of the most unique and appealing properties of DNA for molecular electronics are its double-strand *recognition* and a special structuring that suggests its use for *self-assembly*.

Molecular *recognition* describes the capability of a molecule to form selective bonds with other molecules or with substrates, based on the information stored in the structural features of the interacting partners. Molecular recognition processes may play a key role in molecular devices by: (a) driving the fabrication of devices and integrated circuits from elementary building blocks, (b) incorporating them into supramolecular arrays, (c) allowing for selective operations on given species potentially acting as dopants, and (d) controlling the response to external perturbations represented by interacting partners or applied fields.

Self-assembly, which is the capability of molecules to spontaneously organize themselves in supramolecular aggregates under suitable experimental conditions [25], may drive the design of well-structured systems. Self-organization may occur both in solution and in the solid state through hydrogen-bonding, Van der Waals and dipolar interactions, and by metal-ion coordination between the components. The concept of selectivity approach, which both recognition and self-assembly are based, originates from the concept of information: that is, the capability of selecting among specific configurations reflects the information stored in the structure at the molecular level. It is natural and appealing to use such features to design molecular devices capable of processing information and signals.

By virtue of their recognition and self-assembling properties, DNA molecules seem particularly suitable as the active components for nanoscale electronic devices [26–28]. DNA's natural function of information storage and transmission, through the pairing and stacking characteristics of its constituent bases, stimulates the idea that it can also carry an electrical signal. However, despite the promising development that has been recently achieved in controlling the self-assembly of DNA [29–32] and in coupling molecules to metal contacts [12, 33], there is still a great controversy around the understanding of its electrical behavior and of the mechanisms that might control charge mobility through its structure [34].

The idea that double-stranded DNA, the carrier of genetic information in most living organisms, may function as a conduit for fast electron transport along the axis of its base-pair stack, was first advanced in 1962 [35]. Instead, later low-temperature experiments indicated that radiation-induced conductivity can only be due to highly mobile charge carriers migrating within the frozen water layer surrounding the helix, rather than through the base-pair core [36]. The long-lasting interest of the radiation community [37] in the problem of charge migration in DNA was due to its relevance for the mechanisms of DNA oxidative damage, whose main target is the guanine (Gua) base [38]. Recently, the interest in DNA charge mobility has been revived and extended to other interdisciplinary research communities. In particular, the issue of electron and hole migration in DNA has become a hot topic for a number of chemistry scholars [39, 40] following the reports that photoinduced electron transfer occurred with very high and almost distance-independent rates between donor and acceptor intercalators along a DNA helix [41, 42]. This evidence suggested that double-stranded DNA may exhibit a "wire-like" behavior [43].

From the large body of experimental studies performed in solution that became available in the last decade and appeared in recent reviews [44, 45], several mechanisms were proposed for DNA-mediated charge migration, depending on the energetics of the base sequence and on the overall structural aspects of the system under investigation. These mechanisms include single-step superexchange [41], multistep hole hopping [46], phonon-assisted polaron hopping [47], and polaron drift [48]. The above advances drove the interest in DNA molecules also for nanoelectronics. In this field, by virtue of their sequence-specific recognition properties and related self-assembling capabilities, they might be employed to wire the electronic materials in a programmable way [12, 13]. This research path led to a set of direct electrical transport measurements. In the first reported measurement, μm-scale λ-DNA molecules were found to be "practically insulating" [12]. However, the possibility that double-stranded DNA may function as a one-dimensional conductor for molecular electronic devices has been rekindled by other experiments where, e.g., anisotropic conductivity was found in an aligned DNA cast film [49], and ohmic behavior with high conductivity was found also in a 600-nm-long λ-DNA rope [50].

The above measurements, complemented by other experiments which are discussed in Sect. 2, highlight that, despite the outstanding results that have been recently achieved in controlling the self-assembly of DNA onto inorganic substrates and electrodes, there is currently no unanimous understanding of its electrical behavior and of the mechanisms that might control charge mobility through its structure. Our purpose in this chapter is to review the main experiments that have been performed to measure directly the conductivity of DNA molecules, and to correlate the measurements to the state-of-the-art theoretical understanding of the fundamental electronic and transport features.

1.3 Charge Transport in Device Configuration Versus Charge Transfer in Solution Chemistry Experiments

As already outlined, the interest in the charge migration through DNA grew in three different scientific communities in an almost historical path. The problem originated from the study of genetic mutations related to cancer therapy [37, 38]. It was then reframed in the spirit of determining how fast and how far can charge carriers migrate along the DNA helix in solution [43]. Finally, it was reformulated again in the nanoscience field to question whether such charge motions are capable of inducing large enough currents in DNA-based electronic devices in a dry environment (namely, with the molecules in conditions very different from the native biochemical ones). These research lines proceed separately but bear connections that may finally unravel a uniform vision and interpretation for the mechanisms that control the motion of charge carriers in various DNA molecules. However, care should be taken in advancing a unique paradigm for the interpretation of data coming from different investigation schemes.

Here, we aim at elucidating how the problem is formulated within the "solution chemistry" community and the "solid state" community, and how the experimental investigations are conducted. The theories related to the different classes of measurements are mentioned later in Sect. 3. The theoretical foundations of the relationship between physical observables revealed in solution and in the electrical transport experiments have been recently thoroughly formulated by Nitzan in different regimes for charge mobility (one-step superexchange—tunneling—and multistep hopping) [51].

The experiments in solution, based on electrochemistry techniques, were usually targeted at measuring electron-transfer rates between a donor and an acceptor as a function of the donor–acceptor distance and of the interposed base sequence. The donor is a site along the base stack, where a charge (usually positive, forming a radical cation or "hole") is purposely injected into the structure, and the acceptor is a "hole trapping" site at a given distance. The results are an average signal measured over a large number of molecules. The interpretation is generally given in terms of the change of localization site for the hole. The inherent structure of the molecule is compromised by the transfer process, in the sense that the charge state at distinct sites along the helix before and after the hole migration is different. In these experiments there is no tunneling barrier for the charge to overcome when injected into the molecule.

The experiments in the solid state are based on several techniques, including imaging, spectroscopy, and electrical transport measurements that reveal the electric current flux through the molecule under an external field. The results pertain to single molecules (or bundles) and can be remeasured many times. The roles of the donor and of the acceptor are in this case played either by the metal leads, or by the substrate and a metal tip. The interpretation is generally given in terms of *conductivity*, determined by the electronic energy levels (if the molecular structure supports the existence of localized orbitals and *discrete* energy levels) or bandstructure (if the intramolecular interactions support the formation of delocalized states described by *continuous* energy levels, i.e., dispersive bands). The donor and the acceptor are reservoirs of charges and this fact allows them to leave the charge state along the helix unaltered. It is not specified a priori if the mobile charges are electrons or holes: this depends on the availability of electron states, on their filling, and on the alignment to the Fermi levels of the reservoirs.

In both the indirect electrochemical and the direct transport measurements, the electronic structure of the investigated molecules is important [51]. It determines the occurrence of direct donor–acceptor tunneling or of thermal hopping of elementary charges or polarons. Direct tunneling can occur either "through-space" if the DNA energy levels are not aligned with the initial and final charge sites or reservoirs, or "through-bond" if they are aligned and modulate the height and width of the tunneling barrier. In the case of tunneling, the bridging bases do not offer intermediate residence sites for the moving charges. On the contrary, in the case of thermal coupling and hopping, the moving charges physically reside for a finite relaxation time in intermediate sites at base planes between the donor and the acceptor along their path, although this may cost structural reorganization energy. Whether the inherent DNA electronic structure is constituted of dispersive bands or of discrete levels may be revealed only in the solid-state experiments. In fact, for the motion of individual charges injected into free molecules in solution, probed by electrochemistry tools, it is not important whether such charges find in the molecules a continuum of energy levels or discrete levels available to modulate the tunneling barrier. This is because only the modulation of the tunneling barrier or the donor–bridge–acceptor coupling can be detected. Alternatively, in direct electrical transport measurements, where charges are available in reservoirs (the metal electrodes), it makes a difference if there is a continuum of electron states or discrete levels in the molecular bridge that are available for mobile carriers. For the ideal case of ohmic contacts, a continuum in the molecule will be manifested in smoothly rising current–voltage curves, whereas for discrete levels the measured I–V curves will be step-like revealing quantization.

This chapter is devoted to a review of the latter class of experiments, complemented by the analysis of the theoretical interpretation of the measurements and underlying phenomena.

2
Direct Electrical Transport Measurements in DNA

A series of direct electrical transport measurements *through* DNA molecules that commenced in 1998 was motivated by new technological achievements in the field of electron-beam lithography and scanning-probe microscopy, as well as by encouraging experimental data suggesting high electron-transfer rates. The latter were based on the interpretation of results of charge-transfer experiments conducted on large numbers of very short DNA molecules in solution, in particular by Barton's group at Caltech and by other colleagues [39–47, 52–57]. In a perspective it seems now that care should be taken when projecting from those experiments on the electron-transport properties of various single DNA molecules in different situations and structures, e.g., long vs short, on surfaces vs suspended, in bundles vs single, in various environmental conditions like dry environment, or in other exotic configurations.

Few works have been published since 1998 describing direct electrical transport measurements conducted on single DNA molecules [12, 14, 33, 50, 58–62]. In such measurements one has to bring (at least) two metal electrodes to a physical contact with a single molecule, apply voltage, and measure current (or vice versa). Poor intrinsic conductivity, which seems to be the case for DNA, provides a small measured signal. In such cases the electrode separation should be small, preferably in the range of few to tens of nanometers, yet beyond direct tunneling distance and without any parallel conduction path. The performance of these experiments is highly sophisticated and therefore it is not surprising that the number of the reported investigations is small. Performing good and reliable experiments on single segmented molecules is extremely hard but their interpretation on the basis of the current data is even harder. Not only that, each segmented molecule—a polymer—is intrinsically different from the others in the specific details of its structure. Therefore, also the details of its properties bear some uniqueness. Moreover, the properties of these molecules are sensitive to the environment and environmental conditions, e.g., humidity, buffer composition etc.

Another difficulty that arises in these measurements is that the contacts to a single molecule, as to any other small system, are very important for the transport but hard to perform and nearly impossible to control microscopically. For example, the electrical-coupling strength between the molecule and the electrodes will determine whether a Coulomb blockade effect (weak coupling) or a mixing of energy states between the molecule and the electrodes (strong coupling) is measured. In the case of weak coupling, the size and chemical nature of the molecule between the electrodes will determine the relative contributions of Coulomb blockade phenomena and of the intrinsic energy gap of the molecule to the current–voltage spectra. For the outlined reasons, we find a large variety in the results of the few reported experiments, most of which were done by excellent scientists in leading laboratories.

The question whether DNA is an insulator, a semiconductor or a metal is often raised. This terminology originates from the field of solid-state physics where it refers to the electronic structure of semi-infinite periodic lattices. It is even successfully used to describe the electrical behavior of one-dimensional wires like carbon nanotubes, where a coherent bandstructure is formed. However, it is questionable whether or not this notion describes well, with a similar meaning, the orbital energetics and the electronic transport through one-dimensional soft polymers that are formed of a large number of sequential segments. In these polymers the number of junctions and phase-coherent "islands" is large and may determine the electronic structure and the transport mechanisms along the wire. In some cases it may be those junctions that constitute a bottleneck for the transport. They will determine the overall electric response of the polymer, in spite of suitable energy levels and/or "bands" in the islands that connect those junctions, that could otherwise enable a coherent charge transport. In the case of a strong coupling between the islands along the polymer, a complex combination of the molecular electron states and of the coupling strengths at the junctions will determine the electrical response of the wire.

DNA in particular is sometimes said to be an insulator or a semiconductor. If we assume the possible formation of a long phase-coherent portion, then it may be useful to introduce a distinction between the two terms. In the bulk the difference between a wide bandgap semiconductor and an insulator is mainly quantitative with regard to the resistivity. For DNA and other polymers we may instead introduce the following distinction. If we apply a voltage (even high) across a wide-bandgap polymer and successfully induce charge transport through it *without* changing the polymer structure and its properties *in an irreversible way*, then it would be a wide-bandgap semiconductor. However, if the structure is *permanently damaged* or changed upon this voltage application then it is an insulator. This distinction is important with regard to the relevant experiments, where very high fields are present, and to the methods to check whether the conduction properties of the molecule are reproducible.

In Sects. 2.1 and 2.2 we will review the direct electrical transport experiments reported on DNA single molecules, bundles, and networks.

2.1 Single Molecules

The first direct electrical transport measurement on a single, 16-μm-long λ-DNA molecule was published in 1998 by Braun et al. [12]. In this fascinating experiment the λ-DNA was stretched on a mica surface and connected to two metal electrodes, 12 μm apart. This was accomplished using the double-strand recognition between a short single strand (hang-over) in the end of the long λ-DNA and a complementary single strand that was connected to the metal electrode on each side of the molecule (see Fig. 1). Electrical transport measurements through the single molecule that was placed on the surface yielded no observable current up to 10 V.

Later on in 1999 Fink et al. [50] reported nearly ohmic behavior in λ-DNA molecules with a resistance in the MΩ range. The molecules were a few hundred nanometers long and were stretched across ~2-μm-wide holes in a metal-covered transmission electron microscope (TEM) grid, as shown in Fig. 2. This fantastic technical accomplishment was achieved in a high-vacuum chamber where a holographic image was created with a low-energy electron point source (LEEPS) claimed not to radiatively damage the DNA. Note, however, that the bright parts of the DNA in the images may suggest scattering of the beam electrons from the molecule, which may indicate the presence of scattering points along the DNA that could affect the charge transport along the molecule. The actual measurement was performed between a sharp tungsten tip, which was connected to the stretched molecule

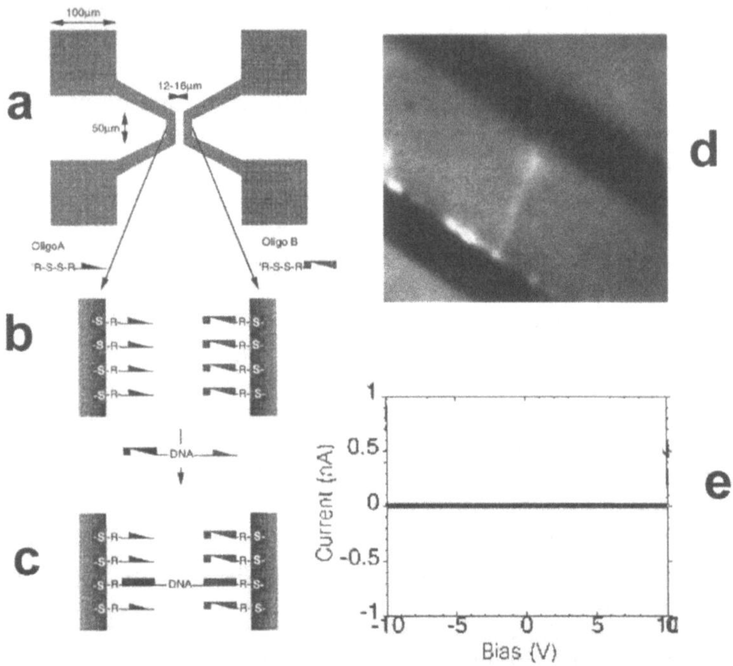

Fig. 1 a–c 16-μm-long λ-DNA was stretched between two metal electrodes using short hang-over single strands complementary to single strands that were preattached to the metal electrodes. **d** A fluorescent image of the DNA molecule, connecting the metal electrodes. **e** The flat, insulating current–voltage that was measured. (from [12], with permission; Copyright 1998 by Nature Macmillan Publishers Ltd)

in the middle of one of the grid holes, and the metal covering the TEM grid. The tungsten tip was aligned using the holographic image. An ohmic behavior was observed in the current–voltage (*I–V*) curves, sustained up to 40 mV and then disappeared. The resistance division between two DNA branches appeared consistent with the ohmic behavior. This result seemed very promising. However, while conduction over long distances was observed later in bundles, it was not repeated in further measurements of single DNA molecules with one exception of a superconducting behavior that is discussed later [60]. The resolution of the LEEPS in this measurement did not enable to determine whether it was a single molecule or a bundle that was suspended between the metal tip and the metal grid.

In a further experiment published in 2000 by Porath et al. [14], electrical transport was measured through 10.4-nm-long (30 base pairs), homogeneous poly(dG)-poly(dC) molecules that were electrostatically trapped [63–64] between two Pt electrodes (see Fig. 3). The measurements were performed at temperatures ranging from room temperature down to 4 K. Current was observed beyond a threshold voltage of 0.5–1 V suggesting that the molecules transported charge carriers. At room temperature in ambient atmosphere, the general shape of the current–voltage curves was preserved for tens of

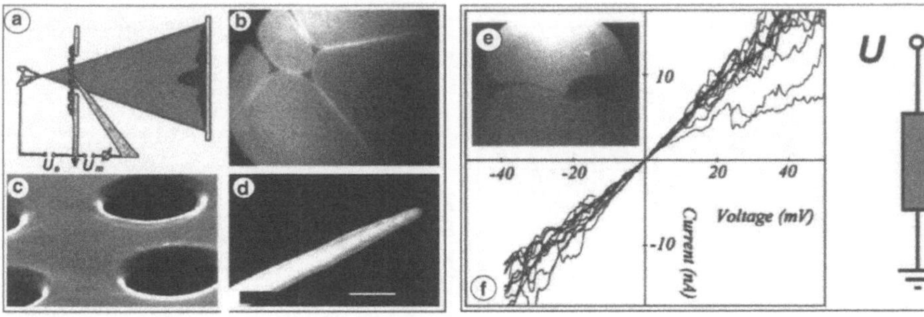

Fig. 2 **a** The LEEPS microscope used to investigate the conductivity of DNA. The atomic-size electron point source is placed close to a sample holder with holes spanned by DNA molecules. Due to the sharpness of the source and its closeness to the sample, a small voltage U_e (20–300 mV) is sufficient to create a spherical low-energy electron wave. The projection image created by the low-energy electrons is observed at a distant detector. Between the sample holder and the detector, a manipulation tip is incorporated. This tip is placed at an electrical potential U_m with respect to the grounded sample holder and is used to mechanically and electrically manipulate the DNA ropes that are stretched over the holes in the sample holder. **b** A projection image of λ-DNA ropes spanning a 2-μm-diameter hole. The kinetic energy of the imaging electrons is 70 eV. **c** SEM image, showing the sample support with its 2-μm-diameter holes. **d** SEM image of the end of a tungsten manipulation tip used to contact the DNA ropes. Scale bar 200 nm. **e** The metal tip is attached to the λ-DNA molecule. **f** *I–V* curves taken for a 600-nm-long DNA rope. In the range of ±20 mV, the curves are linear; above this voltage, large fluctuations are apparent. A resistance of about 2.5 MΩ was derived from the linear dependence at low voltage (from [50], with permission; Copyright 1999 by Nature Macmillan Publishers Ltd)

samples but the details of the curves varied from curve to curve. The possibility of ionic conduction was ruled out by measurements that were performed in vacuum and at low temperature, where no ionic conduction is possible. High reproducibility of the *I–V* curves was obtained at low temperature for tens of measurements on a certain sample, followed by a sudden switching to a different curve shape (see inset of Fig. 4) that was again reproducible (e.g., peak position and height in the *dI/dV* curves, Fig. 4). This variation of the curves in different samples can originate from the individual structural conformation of each single molecule, or from the different formation of the specific contact. The variation of the curves measured on the same sample may be also due to switching of the exact overlap of the wavefunctions that are localized on the bases.

A rather comprehensive set of control experiments helped to verify the results and ensure their validity. The existence of the DNA between the electrodes was verified by incubating the DNA devices with DNase I, an enzyme that specifically cuts DNA (and not any other organic or inorganic material). Following incubation of the sample with the enzyme the electrical signal was suppressed, indicating that the molecule through which the current was measured before is indeed DNA. The procedure was crosschecked by repeating this control experiment in the absence of Mg ions in the enzyme solution

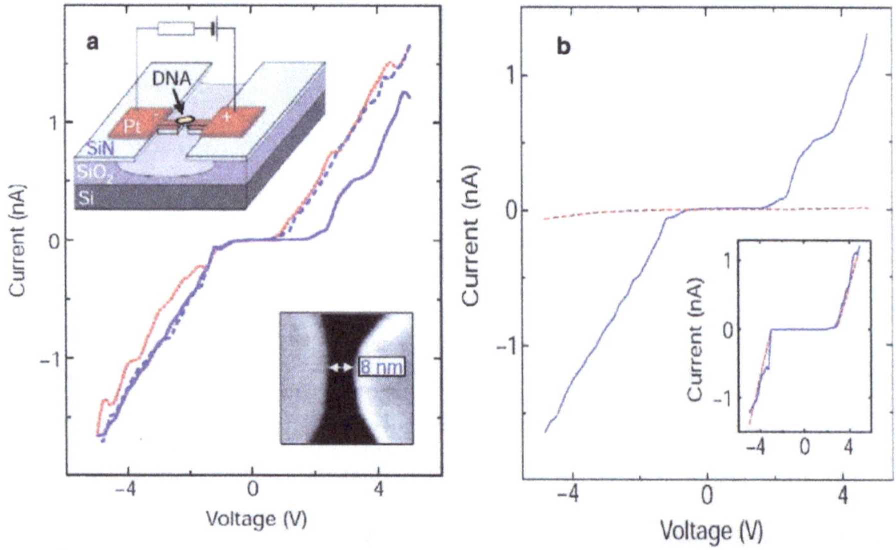

Fig. 3 **a** Current–voltage curves measured at room temperature on a 10.4-nm-long DNA molecule [30 base pairs, double-stranded poly(dG)-poly(dC)] trapped between two metal nanoelectrodes that are 8 nm apart. Subsequent *I–V* curves (different colors) show similar behavior but with a variation of the width of the gap. The *upper inset* shows a schematic of the sample layout. Using electron-beam lithography, a local 30-nm narrow segment in a slit in the SiN layer is created. Underetching the SiO_2 layer leads to two opposite freestanding SiN "fingers" that become the metallic nanoelectrodes after sputtering Pt through a Si mask. The *lower inset* is a SEM image of the two metal electrodes (*light area*) and the 8-nm gap between them (*dark area*). Deposition of a DNA molecule between the electrodes was achieved with electrostatic trapping. A 1-μl droplet of dilute DNA solution is positioned on top of the sample. Subsequently, a voltage of up to 5 V is applied between the electrodes. The electrostatic field polarizes a nearby molecule, which is then attracted to the gap between the electrodes due to the field gradient. When a DNA molecule is trapped and current starts to flow through it, a large part of the voltage drops across a large (2 GΩ) series resistor, which reduces the field between the electrodes and prevents other molecules from being trapped. Trapping of DNA molecules using this method is almost always successful. **b** Current–voltage curves that demonstrate that transport is indeed measured on DNA trapped between the electrodes. The *solid curve* is measured after trapping a DNA molecule as in **a**. The *dashed curve* is measured after incubation of the same sample for 1 h in a solution with 10 mg/ml DNase I enzyme. The clear suppression of the current indicates that the double-stranded DNA was cut by the enzyme. This experiment was carried out for four different samples (including the sample of Fig. 4). The *inset* shows two curves measured in a complementary experiment where the above experiment was repeated but in the absence of the Mg ions that activate the enzyme and in the presence of 10 mM EDTA (ethylenediamine tetraacetic acid) that complexes any residual Mg ions. In this case, the shape of the curve did not change. This observation verifies that the DNA was indeed cut by the enzyme in the original control experiment (from [14], with permission; Copyright 2000 by Nature Macmillan Publishers Ltd)

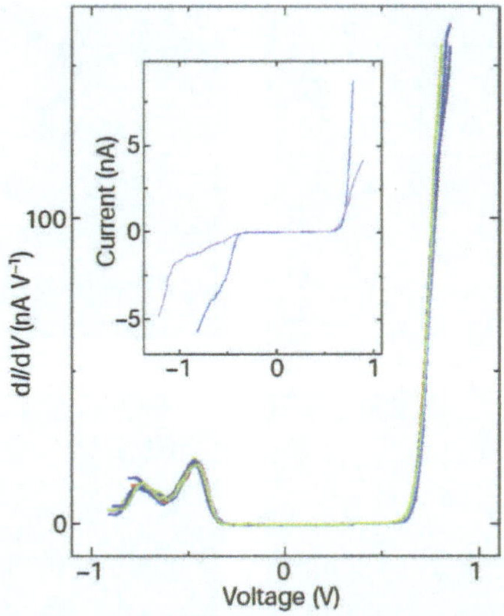

Fig. 4 Differential conductance dI/dV versus applied voltage V at 100 K. The differential conductance manifests a clear peak structure. Good reproducibility can be seen from the six nearly overlapping curves. Peak structures were observed in four samples measured at low temperatures although details were different from sample to sample. Subsequent sets of I–V measurements can show a sudden change, possibly due to conformational changes of the DNA. The *inset* shows an example of two typical I–V curves that were measured before and after such an abrupt change. Switching between stable and reproducible shapes can occur upon an abrupt switch of the voltage or by high current (from [14], with permission; Copyright 2000 by Nature Macmillan Publishers Ltd)

so that the action of the enzyme could not be activated. In this case the signal was not affected by the incubation with the enzyme. This procedure ensured that it was indeed the enzyme that did the cut (see Fig. 3b), thus confirming again that it was the DNA between the electrodes. This experiment clearly proves that short homogeneous DNA molecules are capable of transporting charge carriers over a length at least 10 nm.

Additional experiments were performed in 2001 in the same laboratory by Storm et al. [59], in which longer DNA molecules (>40 nm) with various lengths and sequence compositions were stretched on the surface between planar electrodes in various configurations (see Fig. 5). No current was observed in these experiments suggesting that charge transport through DNA molecules longer than 40 nm on surfaces is blocked.

In parallel, Kasumov et al. [60] reported ohmic behavior of the resistance of λ-DNA molecules deposited on a mica surface and stretched between rhenium–carbon electrodes (see Fig. 6). This behavior was measured at temperatures ranging from room temperature down to 1 K. Below 1 K a particularly unexpected result was observed: proximity-induced superconductivity. The resistance was measured directly with a lock-in technique and no current–voltage curves were presented. This surprising proximity-induced supercon-

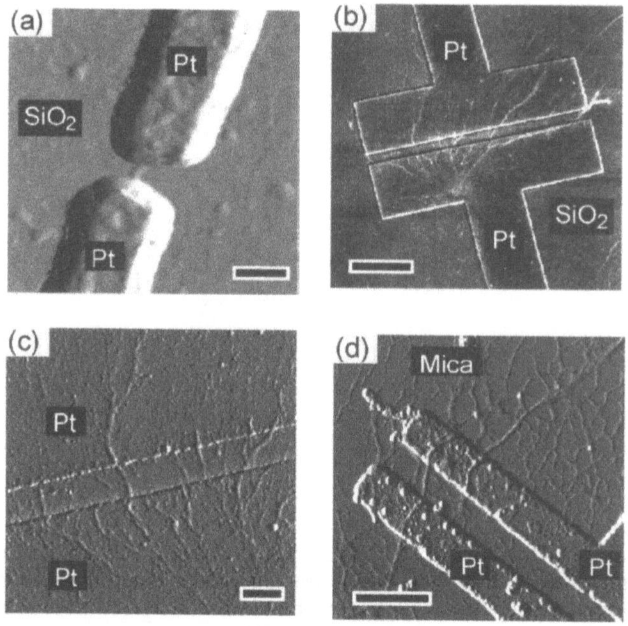

Fig. 5a–d AFM images of DNA assembled in various devices. **a** Mixed-sequence DNA between platinum electrodes spaced by 40 nm. Scale bar 50 nm. **b** Height image of poly(dG)-poly(dC) DNA bundles on platinum electrodes. The distance between electrodes is 200 nm, and the scale bar is 1 μm. **c** High magnification image of the device shown in **b**. Several DNA bundles clearly extend over the two electrodes. Scale bar 200 nm. **d** Poly(dG)-poly(dC) DNA bundles on platinum electrodes fabricated on a mica substrate. Scale bar 500 nm. For all these devices, no conduction was observed (from [59], with permission; Copyright 2001 by the American Institute of Physics)

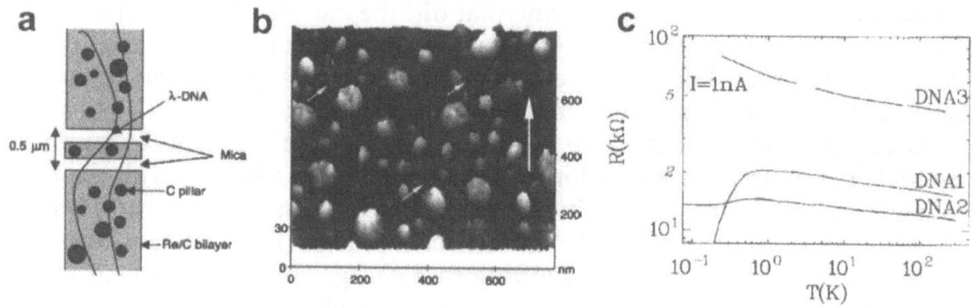

Fig. 6 **a** Schematic drawing of the measured sample, with DNA molecules combed between Re/C electrodes on a mica substrate. **b** AFM image showing DNA molecules combed on the Re/C bilayer. The *large vertical arrow* indicates the direction of the solution flow, used to deposit the DNA. The *small arrows* point towards the combed molecules. Note the forest structure of the carbon film. **c** DC resistance as a function of temperature on a large temperature scale for three different samples, showing the power law behavior down to 1 K (from [60], with permission; Copyright 2001 by Science)

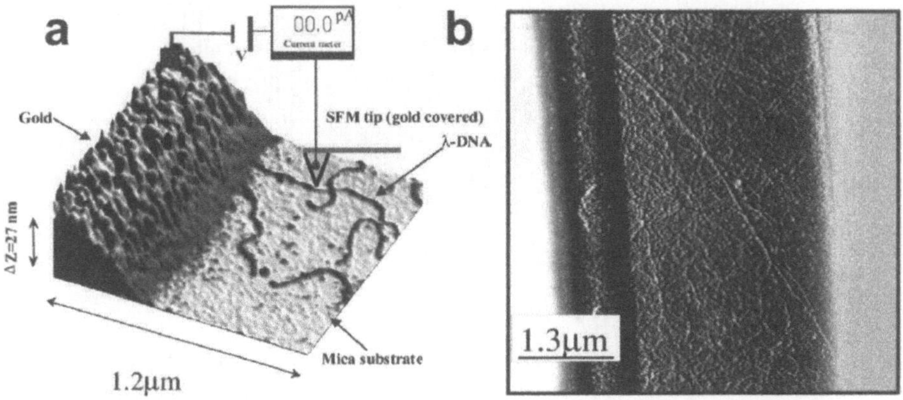

Fig. 7 **a** Three-dimensional SFM image showing two DNA molecules in contact with the gold electrode. The image size is 1.2 μm^2. A scheme of the electrical circuit used to measure the DNA resistivity is also shown. **b** λ-DNA strands connecting two gold electrodes spanned on a bare mica gap. The image analysis leads to the conclusion that at least 1,000 DNA molecules are connecting the electrodes. From the (absence of) current between the electrodes, a lower bound of 10^5 Ωcm per molecule is obtained for the resistivity of DNA at a bias voltage of 10 V (from [58], with permission; Copyright 2000 by the American Physical Society)

ductivity is in contrast to all the other data published so far, and with theory. No similar result was reported later by this or any other group.

In another attempt to resolve the puzzle around the DNA conduction properties, de Pablo et al. [58] applied a different technique to measure single λ-DNA molecules on the surface in ambient conditions. They deposited many DNA molecules on mica, covered some of them partly with gold, and contacted the other end of one of the molecules (>70 nm from the electrode) with a metal AFM tip (see Fig. 7). No current was observed in this measurement. Furthermore, they covered ~1,000 parallel molecules on both ends with metal electrodes (~2 μm apart) and again no current was observed. Yet another negative result, published in 2002, was obtained in a similar experiment by Zhang et al. [33] who stretched many single DNA molecules in parallel between metal electrodes and measured no current upon voltage application. Both results [33, 58] were consistent with the Storm et al. experiment [60].

Beautiful and quite detailed measurements with different results on shorter molecules were reported by Watanabe et al. [61] and Shigematsu et al. [62] using a rather sophisticated technique. A short, single DNA molecule was contacted with a triple-probe AFM. The DNA molecule was laid on the surface and contacted with a triple-probe AFM consisting of 3 CNTs. Two of them, 20 nm apart, were attached to the AFM (see Fig. 8c). In one case, voltage was applied between the nanotube on one side of the molecule and the tip nanotube that contacted the DNA molecule at a certain distance from the side electrode, so that the dependence of the current on the DNA length was

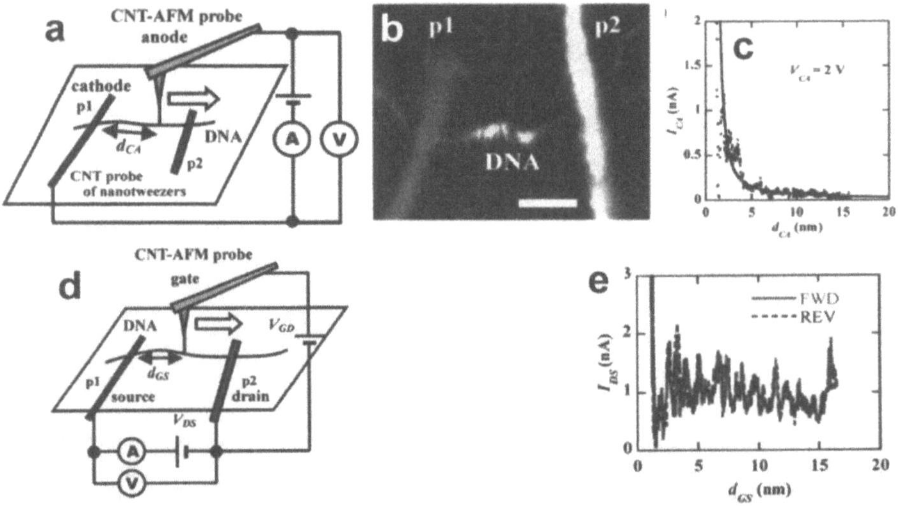

Fig. 8 **a** Schematic of the electric current measurement. Two CNT probes (p1 and p2) of the nanotweezers were set on a DNA. In a two-probe dc measurement, one of the CNT probes (p1) was used as the cathode. A CNT-AFM probe was contacted with the DNA as the anode. The electric current between source and drain was measured while varying the distance between the anode and cathode d_{CA}. **b** AFM image (scale bar, 10 nm) of a single DNA molecule attached with two CNT probes (p1 and p2) of the nanotweezers, which was obtained by scanning the CNT-AFM probe. **c** dependence, d_{CA} of the electric current (I_{CA}) between the electrodes, measured with the electrode configuration shown in **a.** **d** Schematic of the electric measurement under applied gate bias. Two nanotweezer probes (p1 and p2) and the CNT-AFM probe were used as source, drain, and gate electrodes, respectively. The electric current between source and drain was measure with varying the distance between source and gate d_{GS}. **e** d_{GS}-dependence of the electric current (I_{DS}) between the source and the drain electrodes measured with the electrode configuration shown in **d.** The *solid gray line* shows the electric current for the CNT-AFM probe moving from the source electrode to the drain electrode. The *dashed black line* is for the case of the opposite moving direction (from [62], with permission; Copyright 2003 by the American Institute of Physics)

measured under a bias voltage of 2 V between the two electrodes. The current dropped from 2 nA at ~2 nm to less than 0.1 nA in the length range of 6 to 20 nm. In the second experiment reported by this group [62], current was measured between the side nanotubes (20 nm apart) under a bias voltage of 2 V while moving the tip nanotube that served this time as a gate along the DNA molecule. A clear variation of the current due to the effect of the gate electrode, reproducible forwards and backwards, is observed. The current–voltage curves in this experiment are measured through carbon nanotubes. Their conductivity is indeed much higher that that of the DNA molecule and therefore likely to have only a small effect on the *I*–*V*s. However, this and the contacts of the nanotubes to the AFM tip and metal electrodes might still have an effect on the measured results.

From the direct electrical transport measurements on single DNA molecules reported so far one can draw some very interesting conclusions. First, it is possible to transport charge carriers through single DNA molecules, both homogeneous and heterogeneous. This was observed, however, only for short molecules of up to 20 nm in the experiments of Porath et al. [14], Watanabe et al. [61], and Shigematsu et al. [62]. All three experiments demonstrated currents of the order of 1 nA upon application of a voltage of ~1 V. The experiments by Fink et al. [50] and Kasumov et al. [60] showed higher currents and lower resistivities over longer molecules (hundreds of nm), but they were never reconfirmed for individual molecules. In all the other experiments by Braun et al. [12], de Pablo et al. [58], Storm et al. [59], and Zhang et al. [33] that were conducted on long (>40 nm) single DNA molecules attached to surfaces no current was measured. This result is not too surprising if we recall that DNA is a soft, segmented molecule and is therefore likely to have distortions and defects when subjected to the surface force field. This is also manifested in AFM imaging where the measured height of the molecule is different from its "nominal height" [59, 65], partly due to the effect of the pushing tip and partly due to the effect of the surface force field. This force field may be a reason for blocking the current but not necessarily the only one.

The conclusion of poor conductivity in long single molecules on surfaces is further supported by indirect electrostatic force microscope (EFM) measurements, reported by Bockrath et al. [65] and Gómez-Navarro et al. [66]. In these measurements no attraction was found between a voltage-biased metal tip and the λ-DNA molecules lying on the surface. This indicates that the electric field at the tip failed to induce long-range polarization in the molecules on the surface, which would in turn indicate charge mobility along the molecule, as was found for carbon nanotubes.

2.2
Bundles and Networks

A few measurements of direct electrical transport were also performed on single bundles. Other measurements were done on networks formed of either double-stranded DNA [67] or alternative polynucleotides [68]. All the reported measurements showed current flowing through the bundles. We will show a few examples here.

The most productive group in the "networks field" is the group of Tomoji Kawai from Osaka that published an extended series of experiments on different networks and with various doping methods [69, 70, and references therein]. In one of their early experiments they measured the conductivity of a single bundle [67]. This was done in a similar way to the de Pablo experiment [58] (see Fig. 7), placing the bundle between a metal-covered AFM tip on one side of the molecule and under a metal electrode that covered the rest of the bundle (see Fig. 9). The conductivity of a poly(dG)-poly(dC) bundle was measured as a function of length (50–250 nm) and was compared with that of a poly(dA)-poly(dT) bundle. The results showed a very clear

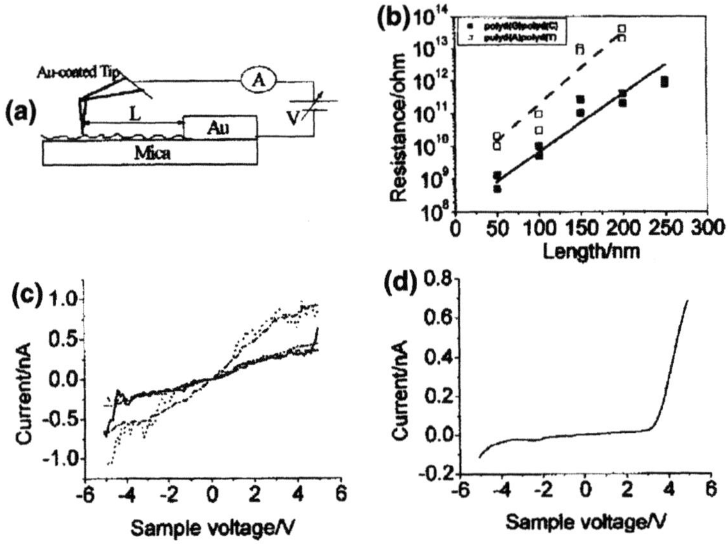

Fig. 9 **a** Schematic illustration of the measurement with a conducting-probe AFM. **b** Relationship between resistance and DNA length for poly(G)-poly(C) (dark marks) for poly(A)-poly(T) (empty marks). The exponential fitting plots of the data are also shown. **c** Typical *I–V* curves of poly(dG)-poly(dC), the linear ohmic behaviors on *L*=100 nm at the repeat measurement of five samples. **d** Rectifying curves of poly(dG)-poly(dC) at *L*=100 nm (from [68], with permission; Copyright 2000 by the American Institute of Physics)

length-dependent conductivity that was about an order of magnitude larger for the poly(dG)-poly(dC) bundle.

One of the interesting measurements among the "bundle experiments" was done by Rakitin et al. [71]. They compared the conductivity of a λ-DNA bundle to that of an M-DNA [71–74] bundle (DNA that contains an additional metal ion in each base pair, developed by the group of Jeremy Lee from Saskatchewan). The actual measurement was performed over a physical gap between two metal electrodes in vacuum (see Fig. 10). Metallic-like behavior was observed for the M-DNA bundle over 15 μm, while for the λ-DNA bundle a gap of ~0.5 V in the *I–V* curve was observed followed by a rise of the current.

Another measurement that follows the line of the Porath et al. [14] experiment was performed by Yoo et al. [75]. In this experiment, long poly(dG)-poly(dC) and poly(dA)-poly(dT) molecules were electrostatically trapped between two planar metal electrodes that were 20 nm apart (see Fig. 11) on a SiO_2 surface, such that they formed a bundle that was ~10 nm wide. A planar gate electrode added another dimension to this measurement. The current–voltage curves showed a clear current flow through the bundle and both temperature and gate dependencies. The resistivity for the poly(dG)-poly(dC) was calculated to be 0.025 Ωcm.

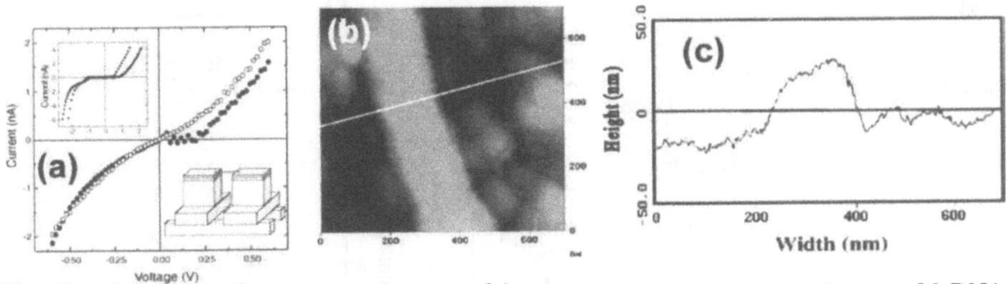

Fig. 10 **a** Current–voltage curves measured in vacuum at room temperature on M-DNA (*empty circles*) and B-DNA (*filled circles*) molecules. The DNA fibers are 15-μm long and the inter-electrode spacing is 10 μm. In contrast to the B-DNA behavior, M-DNA exhibits no plateau in the *I–V* curve. The *lower inset* shows the schematic experimental layout. The *upper inset* shows two representative current–voltage curves measured in vacuum at room temperature on samples of Au–oligomer–B-DNA–oligomer–Au in series. **b** AFM Image of a M-DNA bundle on the surface of the gold electrode (scale bar: 1 μm). **c** Cross section made along the white line in **b** using tapping-mode AFM giving a bundle height of 20–30 nm and width of about 100 nm, which implies it consists of ~300 DNA strands (from [71], with permission; Copyright 2001 by the American Physical Society)

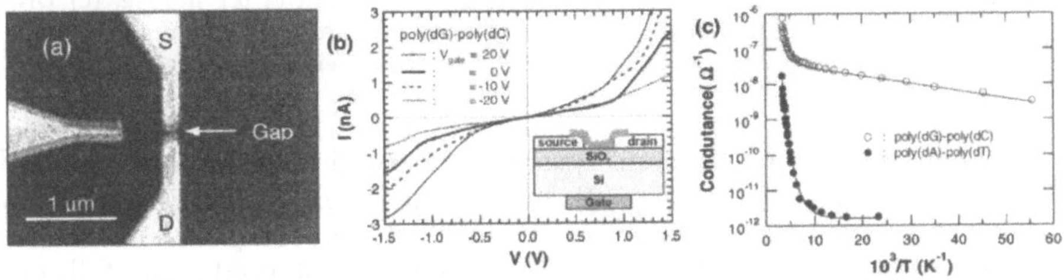

Fig. 11 **a** SEM image of an Au/Ti nanoelectrode with a 20-nm spacing. Three electrodes are shown, S and D stand for source and drain. **b** *I–V* curves measured at room temperature for various values of the gate voltage (V_{gate}) for poly(dG)-poly(dC). The *inset* of **b** is the schematic diagram of electrode arrangement for gate-dependent transport experiments. **c** Conductance versus inverse temperature for poly(dA)-poly(dT) and poly(dG)-poly(dC), where the conductance at V=0 was numerically calculated from the *I–V* curve (from [75], with permission; Copyright 2001 by the American Physical Society)

An interesting experiment on a DNA-based network embedded in a cast film had already been done by Okahata et al. in 1998 [49]. In this pioneering experiment the DNA molecules were embedded (with side groups) in a polymer matrix that was stretched between electrodes (see Fig. 12). It was found that the conductivity parallel to the stretching direction (along the DNA) was ~4.5 orders of magnitude larger than the perpendicular conductivity.

In a recent experiment that was mentioned above with regard to single-molecule measurements, Shigematsu et al. [62] prepared a more complex network that included acceptor molecules. They found a network conductivity that increased with the guanine content.

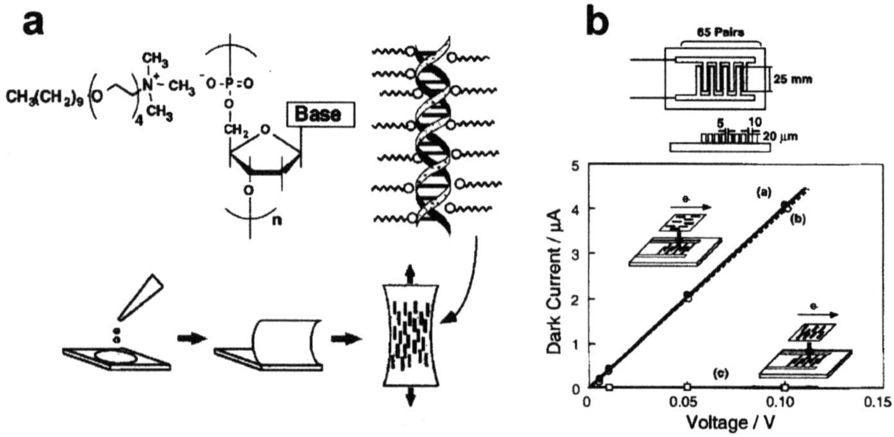

Fig. 12 a Schematic illustration of a flexible, aligned DNA film prepared from casting organic-soluble DNA–lipid complexes with subsequent uniaxial stretching. **b** Experimental geometries and measured dark currents for aligned DNA films (20×10 mm, thickness 30±5 μm) on comb-type electrodes at 25°C. In the dark-current plot, the three curves represent different experimental settings and environments: (a) DNA strands in the film placed perpendicular to the two electrodes (scheme in the *upper inset*) and measured in ambient; (b) the same film as in (a) measured in a vacuum at 0.1 mmHg; (c) DNA strands in the film placed parallel to the two electrodes, both in a vacuum and in ambient (from [49], with permission; Copyright 1998 by the American Chemical Society)

Measurements on a different type of DNA-based material were reported by Rinaldi et al. [68] (see Fig. 13). In this experiment they deposited a few layers of deoxyguanosine ribbons in the gap between two planar metal electrodes, ~100 nm apart. The current–voltage curves showed a gap followed by rise of the current beyond a threshold of a few volts. The curves depended strongly on the concentration of the deoxyguanosine in the solution.

2.3 Conclusions from the Experiments about DNA Conductivity

More and more evidence accumulating from the direct electrical transport measurements shows that it is possible to transport charge carriers along short single DNA molecules, in bundles of molecules, and in networks, although the conductivity is rather poor. This is consistent with the picture that emerges from the electron-transfer experiments. By this picture, that is becoming a consensus, the two most fundamental electron-transfer processes are coherent tunneling over a few base pairs and diffusive thermal hopping over a few nanometers. However, transport through long single DNA molecules (>40 nm) that are attached to the surface is apparently blocked. It may be due to the surface force field that induces many defects in the molecules and blocks the current or any additional reason.

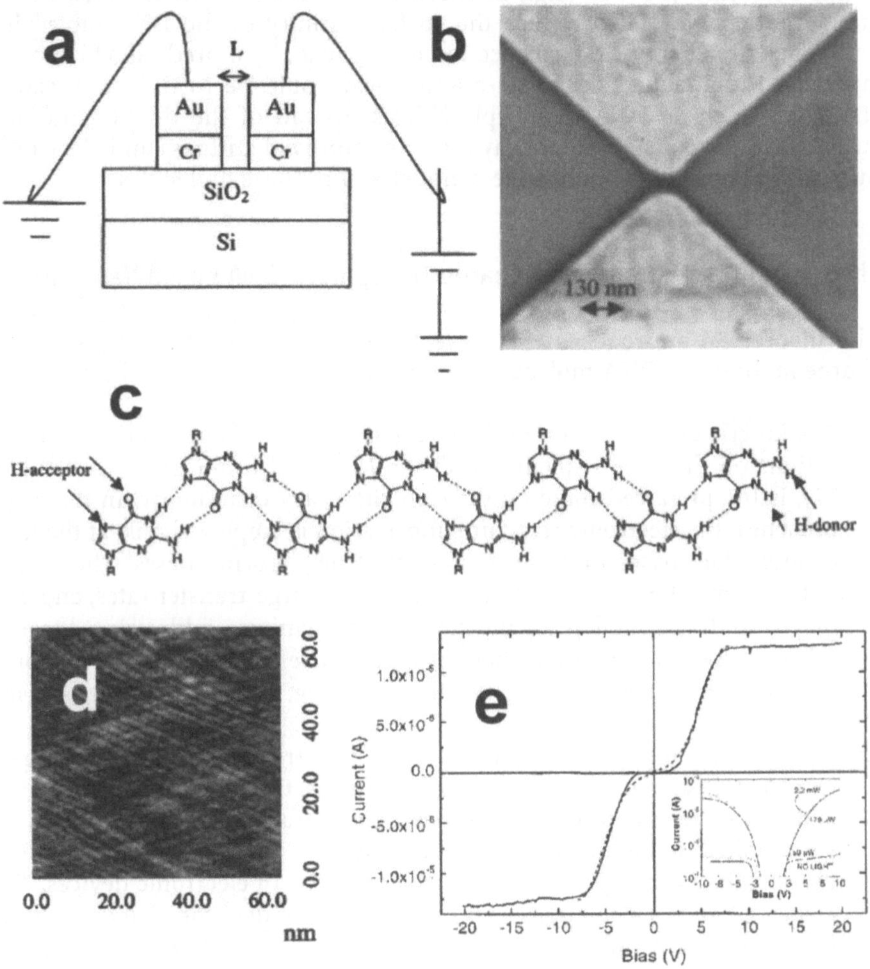

Fig. 13 **a** Schematic of the device used in the experiment. **b** SEM image of the gold nanoelectrodes, fabricated by electron-beam lithography and lift-off onto a SiO_2/Si substrate. **c** Schematics of the ribbon-like structure formed by the deoxyguanosine molecules connected through hydrogen bonds. R is a radical containing the sugar and alkyl chains. **d** AFM micrograph of the ordered deoxyguanosine film obtained after drying the solution in the gap. Regular arrangement of ribbons ranges over a distance of about 100 nm. The ribbon width of about 3 nm is consistent with that determined by X-ray measurements. **e** I–V characteristics of the device (from [68], with permission; Copyright 2001 by the American Institute of Physics)

Therefore, if one indeed wants to use DNA as an electrical molecular wire in nanodevices, or as a model system for studying electrical transport in a single one-dimensional molecular wire, then there are a few possible options. One option is to use doping by one of the methods that are described in the literature [70–74] (e.g., addition of intercalators, metal ions, or O_2 etc.). Another way is to reduce the surface affinity of the DNA molecules and hence the effect of the surface force field (e.g., by a predesigned surface layer) on the attached DNA. Yet another way could be to use more exotic structures such as DNA quadruple helices instead of the double-stranded structure. Such constructions may offer an improved stiffness and electronic overlap that possibly enhance the conductivity of these molecules.

3
Theoretical Understanding of Charge Transport in DNA-based Nanowires

The theoretical approaches that have been applied so far to the study of charge mobility in DNA molecules can be divided into two broad classes.

(i) The kinetic determination of the charge-transfer rates between specific locations on the base sequence, after the Marcus–Hush–Jortner theory [76, 77], is the preferred route by the (bio)chemistry community. In these approaches, the electronic structure information is employed only at the level of individual bases or couples of stacked neighboring bases. The results obtained may be compared to the measured charge-transfer rates, and employed to devise models for the mobility mechanisms, addressing dynamical processes by which the charges might move along the helix, e.g., one-step superexchange, hopping, multiple hopping, polaron hopping [41, 46–48, 78–80].
(ii) The computation of the molecular electronic structure for model and real extended DNA-base aggregates, which affects the quantum conductance and hence the quantities directly measured in transport experiments, is instead linked to the investigations performed by the nanoscience community to explore the role and the efficiency of DNA in electronic devices. The results of such calculations may help devise models for charge mobility from a different point of view, e.g., to unravel the role of the electronic structure in determining the shape of the measured current–voltage characteristics.

The two approaches are not unrelated and a complementary analysis of both kinds of studies would finally shed light onto the detailed mechanisms for charge migration along DNA wires [51]. The kinetic theories are reviewed in other chapters of this book. Here, we focus on results obtained for the electronic structure of extended DNA base stacks, and describe their influence on the electrical conductivity of DNA-based nanostructures.

3.1
Methods to study Quantum Transport at the Molecular Scale

In principle, one would like to perform accurate computations of the relevant measurable quantities to assess the conductivity of the fabricated molecular devices. For coherent transport in the absence of dissipative scattering, the Landauer theory [81–83] is a well-defined frame. It allows one to describe the quantum conductance and the current–voltage characteristics of the system in the device configuration between metallic leads, when the quantum electron structure of the system molecule+leads is known. However, the most manageable formulation of the theory, based on the computation of the Green function (electron propagator), does not allow a straightforward interplay with first-principle methods that are applied to calculate the molecular electronic structure (except for very recent formulations [84–87] that are still very cumbersome and have not yet been applied to DNA-based wires). Therefore, we split our review of the theoretical investigations into two sets. One set is devoted to the parameter-free determination of the electronic structure, without the extension to the measurable quantities (discussed in Sect. 3.2). A second set is devoted to the mesoscopic measurable quantities (such as the I–V curves) with the input electronic structure based on empirical calculations (Sect. 3.3). The former leads to a thorough understanding of the basic physicochemical mechanisms, whereas the latter allows for a direct comparison with the device experiments.

3.1.1
Electronic Structure from First Principles

Among the different possible methods to study the electrical conductivity of solid-state devices, the deepest insight into the process might be gained by studying the energy levels and wavefunctions (or, alternatively for bulk materials, the bandstructure).

The most sophisticated quantum chemistry computational techniques that have been applied to nucleotides are based on the determination of the structure at the Hartree–Fock (HF) level, which includes Coulomb exchange effects but totally neglects correlations. Correlation effects are then taken into account with the application of second-order Møller–Plesset perturbation theory (MP2) to compute relative formation energies [88, 89] with a high degree of accuracy. These cumbersome studies, conventionally named as MP2//HF, provide an accurate determination of the geometry and energetics of stacked and hydrogen-bonded base pairs, but do not presently allow the extension to more complex aggregates. Real nucleotide structures are not accessible to them and require more drastic approximations.

One interesting scheme based on density functional theory (DFT) is particularly appealing, because with the current power of the available computational facilities it enables the study of reasonably extended systems. DFT has been applied with a variety of basis sets (atomic orbitals or plane-waves) and potential formulations (all-electron or pseudopotentials) to complex nucleobase assemblies, including model systems [90–92] and realistic structures [58, 93–95]. DFT [96–98] is in principle an ab initio approach, as well as MP2//HF. However, its implementation in manageable software requires some approximations. The most drastic of all the approximations concerns the exchange-correlation (xc) contribution to the total DFT functional, which is described in a mean-field approach. Whereas the first widely used local density approximation (LDA) functional performs extremely well in bulk contexts, it is not able to quantitatively describe reactive chemistry. Improved generalized-gradient-approximation (GGA) [99–101] and hybrid [102–104] functionals are now able to provide an excellent description of the properties of many-electron systems in molecular environments.

3.1.2
Quantum Transport

In order to obtain estimates of quantum transport at the molecular scale [105], electronic structure calculations must be plugged into a formalism which would eventually lead to observables such as the linear conductance (equilibrium transport) or the current–voltage characteristics (nonequilibrium transport). The directly measurable transport quantities in mesoscopic (and a fortiori molecular) systems, such as the linear conductance, are characterized by a predominance of quantum effects—e.g., phase coherence and confinement in the measured sample. This was first realized by Landauer [81] for a so-called two-terminal configuration, where the sample is sandwiched between two metallic electrodes energetically biased to have a measurable current. Landauer's great intuition was to relate the conductance to an elastic scattering problem and thus to quantum transmission probabilities.

Most implementations of conductance calculations have so far been developed for describing phase-coherent systems, typically semiconductor heterostructures. The latter are fabricated at the micron/submicron scale, a size large enough to justify an approximate treatment of the electronic structure, typically operated by employing a tight-binding (TB) Hamiltonian. However, even with certain classes of smaller and truly molecular systems, an empirical TB treatment of the electronic structure already provides excellent qualitative and in some cases quantitative predictions. This is the case for carbon nanotubes (CNTs), where a simple TB Hamiltonian (including a single π-orbital per carbon atom) is enough to classify a metallic or semiconducting behavior depending on the CNT chirality [107]. In some cases, as in complex structures like DNA wires, the choice of embracing an approximated electronic structure is definitely convenient in order to obtain analytical treatments which might guide the understanding of the basic physics of the system, as Sect. 3.3 presents for the experiment by Porath et al. [14].

3.2
Electronic Structure of Nucleobase Assemblies from First Principles

After briefly presenting some important milestones of MP2//HF studies in the quantum chemistry description of DNA base pairs, we turn to a more extensive discussion of DFT results for extended DNA-base aggregates, including model stacks and real molecular fragments.

3.2.1
Quantum Chemistry

The calculations performed by Šponer and coworkers to evaluate the structure and stability of hydrogen-bonded [88] and stacked [89] base pairs should be retained as important milestones for the application of first-principle computational methods to interacting nucleotides. The authors demonstrated that this kind of theoretical analysis is able to reproduce many of the experimental features, and has a predictive power. Moreover, the description of hydrogen-bonded complexes [88] of DNA bases is propaedeutic to the development of any empirical potential to model DNA molecules and their interaction with drugs and proteins. The conclusions of their investigations may be summarized in the following information: (i) structure of the most favorable hydrogen-bonded and stacked dimers; (ii) rotation and distance dependence of the relative energy of stacked pairs; and (iii) description of the relevant interactions that determine the relative stability of base pairs. Concerning the latter issue, it was found that the energetics of stacked pairs is essentially determined by correlation effects, and therefore can only be accessed through a purely quantum chemical description. On the other hand, the energetics of hydrogen-bonded pairs are well described already at the HF level, and also the DFT treatment is reliable in this context (Van der Waals interactions may be added a posteriori [107]). The investigations by Šponer and coworkers remained limited to the analysis of the structure and energetics of DNA base pairs. The electronic properties were addressed by quantum chemistry methods mainly at the HF level [108–110] (which completely lacks correlation terms): these calculations are discussed in this book in the chapter by Rösch and Voityuk. The notable exception to this restriction is the work performed by Ladik and coworkers [111, 112], who evaluated the shifts of the electron levels and gaps due to correlation effects in the MP2 scheme.

3.2.2
Density Functional Theory

The DFT scheme is more suitable to compute the electronic properties of the extended DNA molecules that are proposed as candidates for electrical wires, and has been successfully applied to a number of different structures. Provided DFT reproduces the main structural features (e.g., bond lengths and angles, stacking distances) in agreement with the MP2//HF calculations, the electronic properties thus derived are fully reliable. DFT is the method of choice for studying reactive chemistry [113].

DFT simulations of DNA-like structures constituted of more than two bases, with an accent on their electronic properties, have become available only recently [58, 90–95, 114]. In our review, we mainly focus on periodic systems obtained by replicating a given elementary unit. For such periodic assemblies, prototypes of DNA-based wires, it is possible to define a "crystal" lattice in one dimension. This allows the extension of the concept of band dispersion and Bloch-like conduction to molecular wires to which one may assign a periodicity length. The advantage of defining a bandstructure for a DNA-based wire is the possible interpretation of experimental results in terms of conventional semiconductor-based device conductivity, using the concept of a band of allowed energy values within which a delocalized electron is mobile. Indeed, we wish to point out that in some recent theoretical studies, the concepts of a bandstructure and dispersive energy bands were ambiguously introduced [115], whereas in principle one could only speak of energy manifolds [58, 93, 94].

In reporting the bandstructure calculations for DNA-based molecular wires, we first focus on the kind of information that may be extracted from the study of model nucleobase assemblies, and then analyze the attempts to treat realistic molecules. Finally, we give a brief account of the environmental effects, e.g., the presence of water molecules and counterions.

3.2.2.1 Model Base Stacks

Di Felice and coworkers performed ab initio calculations of model systems in the frame of plane-wave pseudopotential DFT–LDA(–BLYP) [90]. They considered periodic homoguanine stacks, motivated by the particular role played by this base both in solution chemistry experiments (lowest ionization potential) [52, 53] and in direct conductivity measurements (sequence uniformity, higher stiffness of G-C pairs with respect to A-T pairs) [14]. Their study aimed at understanding the role of various structural features of the base chains in the establishment of continuous orbital channels through the G aggregates. A particular focus was given to: (i) the role of the relative rotation angle between adjacent bases along the π-stack, and (ii) the relative role of π-stacking and hydrogen bonding in structures where both kinds of interactions exist. The results allow one to draw some general conclusions about the configurations that may be conducive to the formation of delocalized wire-like orbitals. Although the model structures are only partially related to G-rich DNA duplexes [14], they are directly related to supramolecular deoxyguanosine fibers [116] that are also suggested as potential building blocks for molecular nanodevices [17, 68].

The π-like nature of the guanine HOMO (see Fig. 14a), namely the protrusion out of the molecular plane, suggests that it might easily hybridize with other similar orbitals in a region of space where a relatively large superposition occurs. The degree of overlap depends on the relative positions of the C–C and C–N bonds where the HOMO charge density is mainly localized, determined by the relative rotation angle of the guanines in consecutive planes of the stack. The calculated interplanar distance in stacked Gua dimers with different relative rotation angles is 3.37 Å, in perfect agreement with experimental and Hartree–Fock data [89]. This value was used to fix the periodicity length in the extended guanine chains. A schematic diagram of the model G stacks is shown in Fig. 14b.

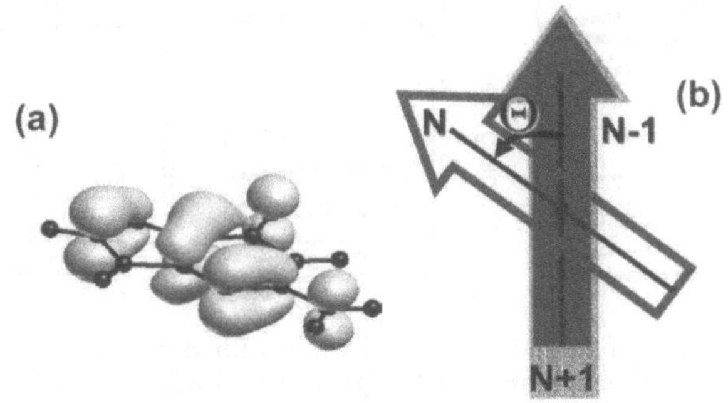

Fig. 14 **a** Isosurface plot of the HOMO of an isolated G molecule, exhibiting a clear π character. **b** Pictorial illustration of the construction of a model periodic wire through the stacking of G bases. To model the periodic structures, Di Felice et al. [90] employed a unit supercell containing two G molecules, including only nearest-neighbor interactions. The models do not reproduce the continuous helicity characteristic of a G strand in DNA. The helicity along the axis of the pile goes instead back and forth at successive steps: if the rotation between planes N−1 and N is Θ, that between planes N and N+1 is $-\Theta$, and plane N+1 is equivalent to N−1. Since nearest-neighbor coupling is already very weak and configuration-dependent, it is expected that longer-range interactions would not contribute effectively to orbital delocalization

Among the several relative rotation angles that were considered, the configurations most representative for the discussion about a viable band-like conductivity mechanism in guanine π-stacks are illustrated in Fig. 15 (insets), along with the computed bandstructure diagrams. The conclusion that can be drawn from the bandstructure analysis of these model guanine strands is that dispersive bands may be induced only by a large spatial π overlap of the HOMO (LUMO) orbitals of adjacent bases in the periodic stack. Such an overlap is maximum for eclipsed guanines (Fig. 15, left), and very small for guanines rotated by 36° (Fig. 15, right) as in B-DNA. These results suggested that a band-like conductivity mechanism, occurring via band dispersion and almost free-like mobile carriers (which should be injected through a suitable doping mechanism), is not viable along frozen G-rich stacks. It cannot be excluded that atomic fluctuations locally induce an amount of overlap larger than in frozen B-DNA, with partial interaction and bandstructure formation at least over a typical coherence length. This is possible for a short length, whereas other dynamical mechanisms should be invoked to explain long-range charge migration. As a final remark, we note that the band dispersion found in the model guanine chains described in this subsection was solely due to the translational symmetry in the infinite wire, not including the helical symmetry. This remark and possible ambiguities about the way different authors report their results will be clarified in the following subsection.

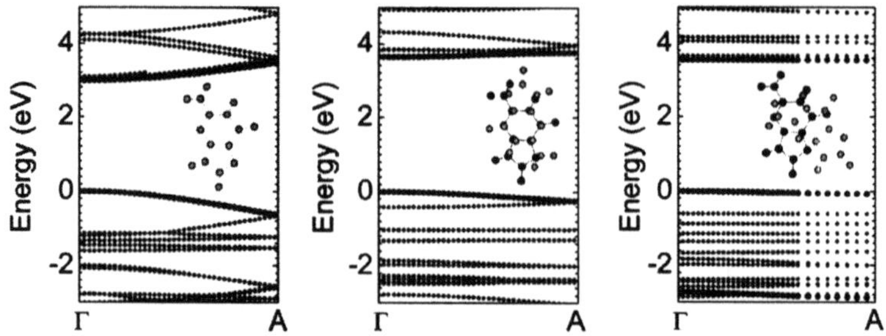

Fig. 15 Bandstructure of periodic model G-wires with different angles Θ (defined in Fig. 14), computed between Γ and the A edge of the 1D BZ. The bandstructure for each system was calculated at several k points of the 1D BZ along the stack axis, interpolating from the self-consistent charge density obtained at one special k point. The *insets* show the top-view geometries of the selected periodic stacks: *gray and black spheres* indicate atoms on two consecutive planes. *Left*: Θ=0°, the ΓA dispersion of the HOMO-derived (LUMO-derived) band is 0.65 eV downwards (0.52 eV upwards). *Middle*: Θ=180°, the ΓA dispersion of the HOMO-derived (LUMO-derived) band is 0.26 eV downwards (0.13 eV upwards). *Right*: Θ=36°, the ΓA dispersion for the HOMO- and LUMO-derived bands is vanishing, index of poor interaction. For Θ=36° the π overlap between consecutive stacked planes is negligible and the energy levels remain flat and degenerate. The π overlap becomes larger for Θ=180° and is maximum for Θ=0°: this is reflected in an increasing band dispersion for both the valence and conduction bands. The effective masses obtained by a parabolic interpolation of the top valence and bottom conduction bands are in the range of 1–2 free electron masses for Θ=0° and Θ=180° (typical of wide-bandgap semiconductors). The DFT energy gap (underestimated [96]) is in the range of 3.0–3.5 eV (adapted from [90], with permission; Copyright 2002 by the American Physical Society)

3.2.2.2
Realistic DNA-Based Nanowires

From the above results, it appears very unlikely that even continuous and uniform base sequences form true semiconducting bands (and associated delocalized orbitals along the helix axis). Nevertheless, there might be other intervening mechanisms of orbital mixing, characteristic of supramolecular structures and without an exact equivalent in the solid-state inorganic crystals, that may induce the effective behavior of a semiconductor. We illustrate this idea with a few examples that appeared in the literature [58, 93, 115], concerning homo-Gua [93] and Gua-Cyt [58, 115] aggregates that resemble realistic structures and include both π-stacking and hydrogen-bonding interactions. The mechanism suggested for a semiconducting behavior, alternative to pure crystal-like Bloch conductivity, involves manifolds of localized levels: These manifolds are formed as a consequence of the interbase interactions that do not involve chemical bonding.

DFT simulations of periodic wires show that the weak coupling between the building blocks (Gua, or Gua-Cyt pairs) contained in the periodicity length split the energy levels of the coupled orbitals, which should be otherwise degenerate in the absence of interplanar interaction. Such a splitting results in the appearance of a "band" of closely spaced energy levels. Although each level is nondispersive, the complete set of similar orbitals (e.g., HOMO) is gathered into a band with a given amplitude, which plays the same role as a dispersive band if the energy splitting between levels in the manifold is small. The formation of energy manifolds was found in both poly(G)-poly(C) [58, 115] and G-quartet [93] wires, for both occupied and empty levels, with amplitudes dependent on the particular molecule and on the computational method. We discuss in the following some quantitative features of the "manifold mechanism" for the origin of a semiconducting bandstructure.

De Pablo and coworkers [58] performed linear-scaling pseudopotential numerical-atomic-orbital DFT–PBE [117] calculations of poly(dG)-poly(dC) DNA sequences with periodicity length corresponding to 11 base planes, in dry conditions [114]. The electronic structure was determined for the optimized geometry. The ordered poly(dG)-poly(dC) wire was characterized by filled and empty "bands" around the Fermi level constituted of 11 states, i.e., one state for each base pair. The highest filled band was derived from the HOMOs localized on the G bases and had a bandwidth of 40 meV (effective dispersion for hole conductivity). The lowest empty band was derived from the LUMOs localized on the C bases and had a bandwidth of 250 meV (effective dispersion for electron conductivity), meaning an effective mass in the range typical of wide-bandgap semiconductors. The LUMO-derived band was separated by an energy gap of 2 eV from the HOMO-derived band: this value is affected by the well-known underestimate of DFT-computed energy gaps [96]. Artacho and coworkers described the establishment of a "band" in terms of the helical symmetry [114]. By virtue of this symmetry, the manifold of energy levels of the Bloch orbitals found at the Γ point of the 1D BZ is split into equally spaced reciprocal-space points along the helix axis, giving an effective band dispersion. The effective electron orbitals (shown as isosurfaces in the original work [58]) are obtained as linear combinations of the one-particle computed Bloch orbitals. The authors also noticed that the wide bandgap itself does not rule out electrical conduction, if any doping mechanisms capable of injecting free carriers (e.g., defects in the hydrogen atoms or counterions saturating the phosphates) are active in the molecule. In the same work [58, 114], it was shown that a defected poly(dG)-poly(dC) sequence exhibits electronic localization over few base pairs, with consequent coherence breaking and exponential decay of the conductance with length.

A similar behavior was found for a nonperiodic 20-base-pair-long poly(G)-poly(C) molecule, by means of valence-effective Hamiltonian (VEH) calculations, whose accuracy is claimed comparable to that of DFT [115]. In this investigation the DNA double helix was modeled by two separate strands, assuming that the hydrogen bonding in each plane gives a "weak" contribution to the Gua-Cyt interaction. We wish to point out that the hydrogen-bonding contribution to the energetics is not weak, but stronger than the base-stacking contribution [88, 90]. A more suitable rephrasing of the concept requires the specification that the weakness of hydrogen bonding is limited to its contribution to interbase orbital hybridization and delocalization [58, 90–91]. Additionally, for the finite poly(G)-poly(C) molecule investigated by Hjort and Stafström the HOMO-derived bandwidth was found to be 0.2 eV. This value is much larger than that found for the periodic wire investigated by de Pablo and coworkers [58], and was not clearly discussed in terms of a manifold of localized orbitals, thus giving the ambiguous interpretation of a real dispersive Bloch-like band.

The semiconducting-like behavior induced by split-level effective bands was recently identified for a G-quartet nanowire [93] which is suggested as an improved structure for an electrical molecular wire. Tubular sequences of G tetramers were investigated by means of pseudopotential plane-wave DFT–BLYP calculations of the equilibrium geometry and electronic structure. In the same way as the base pairs (G-C and A-T) stack on top of each other to form the inner core of double-helix DNA, the tetramers (G4) are the building blocks of a quadruple-helix form of DNA, labeled as G4-DNA. Each tetrameric unit is a planar aggregate of hydrogen-bonded guanines arranged in a square-like configuration (see Fig. 16a) with a diameter of 2.3 nm, slightly larger than the 2.1 nm of native DNA. By stacking on top of each other as shown pictorially in Fig. 16b, these G4-DNA planes form a periodic columnar phase with a central cavity that easily accommodates metal ions coordinating the carbonyl oxygen atoms. The G4-DNA quadruple helix was simulated by periodically repeated supercells, containing three stacked G4 tetramers, separated by 3.4 Å and rotated by 30° along the stacking direction (Fig. 16b,c). The starting atomic configuration was extracted from the X-ray structure of the *G-quadruplex d(TG4T)* [119].

The electronic bandstructure of the K^+-filled G4 quadruple helix is shown in Fig. 17 (left), along with the total DOS (right). The special symmetry of the G4 quadruple helix increases the spatial overlap between consecutive planes with respect to a segment of G-rich B-DNA, thus suggesting (after the discussion above) a possible enhancement of the band-like behavior. However, the results presented in Fig. 17 reveal a different situation. It is found that the interplane π superposition is not sufficient to induce the formation of delocalized Bloch orbitals and dispersive energy bands. The bandstructure shows in fact that the bands remain flat, typical of supramolecular systems in which the electron states are localized at the individual molecules of the assembly. Nevertheless, another mechanism for delocalization takes place. The plot in Fig. 17 (left) identifies the presence of multiplets (or manifolds), each constituted of 12 energy levels. The 12 electron orbitals associated with a multiplet have identical character and are localized on the 12 guanines in the periodic unit cell. The energy levels in a multiplet are separated by an average energy difference of 0.02 eV, smaller than the room temperature energy K_BT: therefore, the coupling with the thermal bath may be sufficient to mix the G-localized orbitals and produce effective delocalization. The resulting DOS in Fig. 17 (right) shows that the multiplet splitting effectively induces the formation of dispersive energy peaks. Filled and empty bands are separated by a bandgap of 3.5 eV. The most relevant DOS peak for transport properties is π-like. It is derived from the HOMO manifold, and has a band-

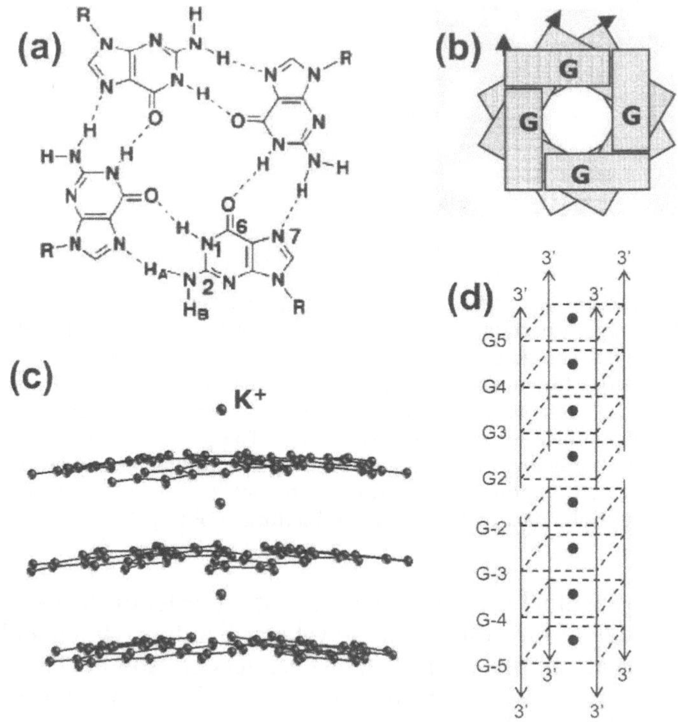

Fig. 16a–d Inner-core structure of the computed G4-wires. **a** A planar tetrad G4, which constitutes the basic building block for stacking planes. **b** Pictorial illustration of the rotation between adjacent stacked planes: each plane is represented as a *square* and in each step along the quadruple helix the guanines are rotated by 30 degrees. **c** Atomic model of the periodicity unit in the computed G4-wires. **d** Pictorial illustration of the G4-helix molecules in the crystal structure. The crystallized quadruplexes [118] are finite molecules containing eight G4 planes, separated in two sets of four G4 planes: within one set (e.g., upper four planes in panel **d**), consecutive planes are connected via a sugar-phosphate backbone similar to that in DNA, whereas the backbone is interrupted and the polarity is inverted at the interface between two different sets (see the break between the upper and lower four planes in panel **d**). In the computational work by Calzolari et al. [93], a periodic infinite (1D crystal-like) wire was constructed from three (see panel **c**) of the four planes in one of the two sets, based on the evidence that the fourth plane in the stack is equivalent to the first one by symmetry. The external backbone was neglected in the simulations. In the central cavity of the tubular sequence, one K^+ ion was inserted between any two planes, as suggested from the X-ray analysis (Na^+ ions were present in the crystals) (adapted from [93], with permission; Copyright 2002 by the American Institute of Physics)

width of 0.3 eV. Although the valence bands do not form a continuum, they exhibit dispersive peaks in precise energy ranges, which may host conductive channels for electron/hole motion.

The manifold energy splitting is accompanied by the formation of delocalized orbitals as shown by the contour and isosurface plots in Fig. 18: a

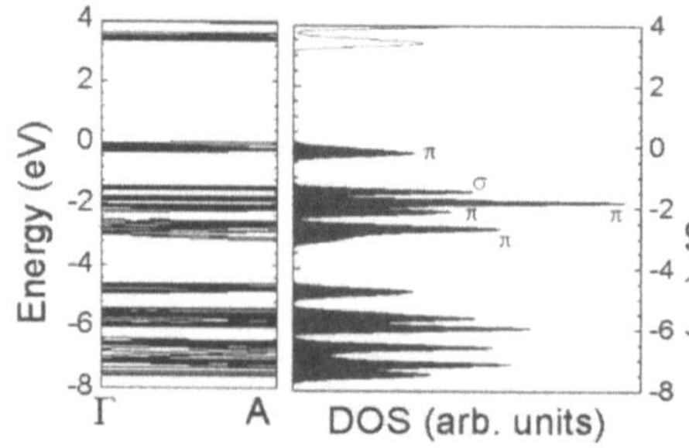

Fig. 17 Bandstructure (*left*) and DOS (*right*) computed for the G4-wire. In the DOS, the *shaded* (*white*) *region* pertains to occupied (empty) electron states. The labels π and σ identify the orbital character. Although the bandstructure obtained with translational symmetry identifies the localized character of the one-particle electron wavefunctions, the calculated DOS is consistent with a model of a semiconducting nanowire, which hosts extended channels for electron/hole motion. The peaks in the effective band-like DOS result from the energy spreading of the single-molecule energy levels that group into manifolds (see the groups of flat bands in the *left panel*) (adapted from [93], with permission; Copyright 2002 by the American Institute of Physics)

clear mobility channel is identified at the outer border of the G4 column. In the ground state of the system, all the valence bands are filled and the conduction bands are empty, so that mobile carriers are absent. Therefore, efficient doping mechanisms, that may eventually rely on the native structural properties of these G4-DNA wires and on the presence of cations, should be devised in order to exploit them as electrical conductors. Indeed, we note that in the study of the G4-DNA-like wires [94] the K semicore states were not taken into account, and that they may be able to provide hybridization with the base stack and an intrinsic doping factor. Such developments move along the direction of investigating the electronic modifications introduced in DNA helices by metal cations inserted in the inner core [71–73, 119–120].

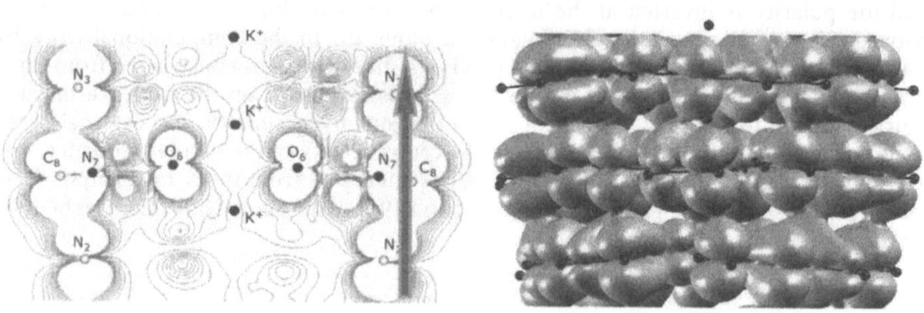

Fig. 18 Contour (*left*) and isosurface (*right*) plots of the HOMO-band convolution of electron states for the G4-DNA base-core structure [93]. The delocalized character along the guanines is evident, as highlighted by the *arrow* parallel to the axis of the stack (adapted from [93], with permission; Copyright 2002 by the American Institute of Physics)

3.2.2.3
Effects of Counterions and Solvation Shell

Two recent DFT calculations, performed at the upper limits of the computational power available with the most sophisticated parallel computers, for an infinite wire [94] and for a finite four-base-pair molecule [95], have addressed the static and dynamic role of counterions in the determination of the electron energy levels and wavefunctions.

Gervasio and coworkers analyzed a periodic nucleotide structure obtained from the finite molecule d(G*p*C*p*)$_6$ in the Z-DNA conformation. The crystal structure of this molecule is known and was assumed as the starting configuration for ab initio molecular dynamics/quenching simulations, with 1,194 atoms in the periodically repeated supercell including the sugar-phosphate backbone, water molecules, and Na^+ counterions [94]. The positions of the latter, not resolved in the crystal structure, were initially assigned on the basis of considerations about the available volume along the backbone and grooves. All the atoms were then relaxed. Besides interesting results about novel arrangements of water clusters surrounding and penetrating the double helix, the authors report the analysis of the electronic structure. The outstanding outcome of this computation is the evidence for Na^+-related electron states within the bandgap occurring between Gua- and Cyt-levels (Fig. 19), which would pertain to the electronic structure of the base stack alone. As a critical comment to this work, we point out the following considerations: (i) the description in terms of Gua and Cyt manifolds is in agreement with the calculations described in the previous sec-

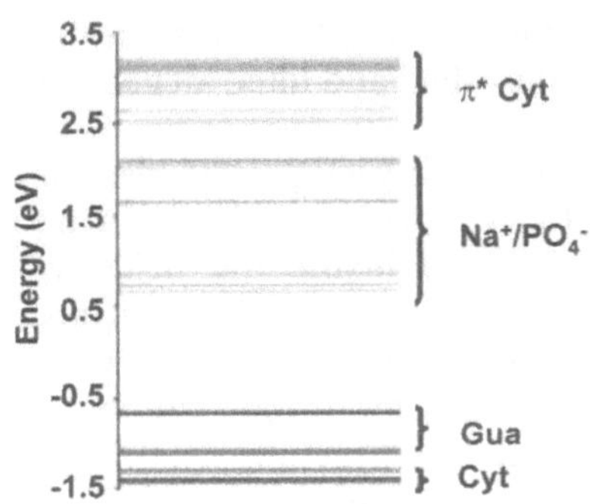

Fig. 19 Schematic level diagram around the Fermi level, for the infinite DNA wire studied by Gervasio et al. [94]. The Fermi level positioned in the *middle* of the gap was chosen as the zero of energy. The highest occupied "band" is constituted of a manifold of 12 states localized on the 12 Gua bases contained in the periodic unit and originated from the Gua-HOMO. Cyt-localized states (π^* Cyt) appear as another manifold at 3.16 eV above the HOMO-band. However, empty electron states due to metal counterions and phosphates are revealed at 1.28 eV above the HOMO-band (see also [92]), and the ground-to-excited-state transitions are therefore related to charge transfer between the inner and outer helix (adapted from [94] with permission; Copyright 2002 by the American Physical Society)

tions; and (ii) the effect of the static metal cations does not destroy the Gua-Cyt underlying "bandstructure", but only introduces additional empty electron states, not hybridizing with the Gua and Cyt orbitals, that might be appealing for doping mechanisms. Therefore, the studies that address the electronic structure of the base core stack remain valid as a fundamental point to understand more complex mechanisms that arise by complicating the geometry.

Another interesting account of Na^+ counterions was devoted to their dynamical role [96]. Barnett and coworkers performed classical molecular dynamics simulations of a finite B-DNA duplex d(5′-GAGG-3′) with an intact sugar-phosphate backbone, including the neutralizing Na^+ counterions and a hydration shell. The classical calculations allowed them to sample the Na^+ "visitation map" (i.e., the map of the sites explored by the Na^+ counterions during the dynamical evolution), from which selected configurations differing for the positions of the cations (populating either backbone sites or helix grooves) were extracted and described by DFT quantum calculations. From their results, the authors identify an "ion-gated transport" mechanism. This mechanism is based on the fact that the hole, described as a total-charge difference between the charged and the neutral system, becomes localized at different core sites depending on the cation positions. Therefore, by migrating along the molecule axis outside the helix, the metal cations drive the hole hopping between G bases and GG traps. Differently from the other studies described in Sect. 3.2, this latter research is not dedicated to the investigation of band-conduction channels through the establishment of delocalized electron states, but addresses the issue of hole localization at various sites induced by structural fluctuations. We believe that both points of view are of relevance for dc conductivity measurements in the solid state, where in principle ionic motions are more frozen than in solution. To which extent this is true has not yet been established and likely depends on the experimental settings.

Finally, we note that the investigations described in this subsection address the role of cations external to the DNA helix. To our knowledge, theoretical studies of cations inside the helix, possibly modifying the base pairing through electronic hybridization, are still in their infancy. Two notable examples along this way are the case of Pt ions interacting with Ade-Thy base pairs [121] modifying the hydrogen-bonding architecture, and that reviewed in the previous section of K ions inside the G4 quadruple helix [93]. These metallized DNA structures deserve special attention because they are lately becoming of interest as metallic nanowires [71–73, 119–120].

3.3
Evaluation of Transport Through DNA Wires Based on Model Hamiltonians

The available first-principle calculations of the electronic properties of DNA molecules, reviewed in the previous section, are complemented by the so-called model Hamiltonian studies [122–125]. The latter typically grasp *partial* aspects of the targeted physical system since they are approximate and are not parameter-free theories: parameters are typically fixed by comparison with experiments or with more complex theories such as DFT. On the other hand, model Hamiltonians possess the valuable potential of gaining intuition on the physical mechanisms of the system at hand due to the complexity reduction that they apply. In most cases they provide analytical solutions and allow one to control the outcoming physics in the parameter space. Moreover, additional physical effects spanning phonon coupling, electronic correlations, and external driving fields might be added in a modular way.

3.3.1 *Scattering Approach and Tight-Binding Models*

The recent progress in nanofabrication unveiled to the experimental investigation the transport properties of structures from the mesoscopic to the molecular scale. Before this was possible, it was already clear that below a certain critical size, the classical behavior of the conductance—scaling proportional to the sample transversal area and inversely proportional to the sample length (Ohm's law)—would break down [126]. The critical size is dominated by the so-called phase-coherence length, that is the length over which quantum coherence is preserved, thus representing a boundary between macroscopic and mesoscopic systems. Moreover, for samples smaller than the electron mean free path (in the so-called ballistic regime) the quantum mechanical wave nature of the electron being transported through the sample turns the classical conduction issue into a quantum wave scattering problem. This idea received its first satisfactory description in 1957 by Landauer [81] and is also known as the "scattering approach". Landauer's theory is now the privileged tool to describe transport through molecular devices and in particular through dry DNA wires.

The basic components of the scattering approach are the scatterer itself (in our case a sample of molecular size) contacted by two external macroscopic metallic electrodes (which we can conventionally call left and right leads). Figure 20 illustrates a typical two-terminal device. The leads are represented by a well-defined electrochemical potential, the Fermi level E_F, providing a reservoir of electrons at the thermal equilibrium. By slightly biasing the electrochemical potential of one of the two leads with respect to the other, a net current can possibly flow through the sample. At a low bias, the ratio between the current and the bias voltage defines the so-called linear conductance g. Following Landauer's intuition, this linear conductance at zero temperature is proportional to the quantum mechanical transmission T, by a factor $G_Q=2e^2/h=(12{,}906\ \Omega)^{-1}$, which is the quantum-of-conductance. This corresponds to writing $g=G_QT(E_F)$ for electrons injected from a lead with a fixed electrochemical potential E_F. The transmission $T(E)$ can be deduced from the scattering matrix (S-matrix) and ul-

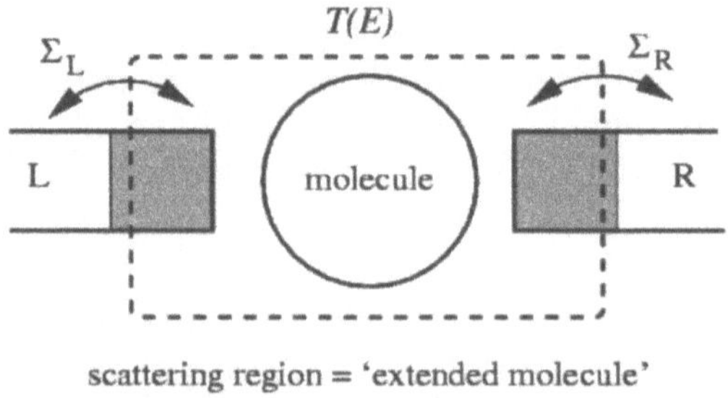

Fig. 20 Schematic representation of a two-terminal device. The scattering region (enclosed in the *dashed-line frame*) with transmission probability $T(E)$ is connected to semi-infinite left (L) and right (R) leads which end in electronic reservoirs (not shown) at chemical potentials E_L, and E_R, kept fixed at the same value E_F for linear transport. By applying a small potential difference electronic transport will occur. The scattering region or molecule may include in general parts of the leads (*shaded areas*) (adapted from [105] with permission; Copyright 2002 by Springer)

timately from the Hamiltonian of the system after the Fisher–Lee relation [127][1].

It is worth noting that both the Landauer formula and the Fisher–Lee relation are valid in the rather idealized regime in which inelastic scattering through the scattering region can be neglected [82]. Given this idealization, and very close to thermal equilibrium (small voltage drops between the left E_L and the right E_R electrochemical potentials), the current at zero temperature can be estimated as

$$I[V = (E_R/e) - (E_L/e)] = \int_{E_L/e}^{E_R/e} T(E)\, dE. \tag{a}$$

For estimates of the current in cases that are truly out-of-equilibrium, the formalism which we have introduced must be generalized to accommodate phase correlations (Keldysh formalism) [128]. Such modeling developments are strongly demanded in the near future, to evaluate charge currents through molecular wires under a high bias, given the wide voltage range in the experiments.

Within the Landauer approach, the computation of the quantum conductance is thus traced back to the knowledge of the electronic structure—i.e., the Hamiltonian—of the target system "molecule+leads". The best-developed implementations of the Landauer framework employ tight-binding Hamilto-

[1] The key ingredient here is the scatterer Green function, which is the mathematical tool implementing the quantum propagation of an electron through the sample. It is defined as the inverse operator of the scatterer Hamiltonian H_{sc} via the relation $G_{sc}(E)=(E+i0^+-H_{sc})^{-1}$, and the so-called (left and right) lead self energies $\sum_L$ and $\sum_R$, which are again obtainable from the Green function of the leads. The Fisher–Lee relation [127] finally reads as $T(E)=4\,Tr\{\Delta_L(E)G(E)\Delta_R(E)G^{\dagger}(E)\}$. G is the dressed scatterer Green function (dressed by the presence of the leads) given by $G^{-1}=G^{-1}{}_{sc}+\sum_L+\sum_R$. Δ_α (α=L,R) is the lead spectral density $\Delta_\alpha=i/2[\sum_\alpha(E+i0^+)-\sum^+_\alpha(E+i0^+)]$.

nians based on localized orbitals providing the simplest guess for the electronic structure of a molecular system, with parameter optimization from experiments or ab initio calculations. TB models provide an ideal platform for the computation of the Green function on which the transmission coefficients depend [82]. A generic TB Hamiltonian, to describe a molecular system at hand, is typically written in an orthogonal basis $\{b_{i,\sigma}\}$ that comprises a single maximally compact orbital per atom (taking into account the spin coordinate σ). In a supramolecular system such as a DNA wire, one orbital per elementary molecule (e.g., the G and C bases in the following example) is included in the basis set. In the TB basis the molecular Hamiltonian is

$$H_{\text{mol}} = \sum_{i,\sigma} \varepsilon_i b_{i\sigma}^{\dagger} b_{i\sigma} - \sum_{\langle i,j \rangle,\sigma} t_{ij} b_{i\sigma}^{\dagger} b_{j\sigma}, \tag{b}$$

where ε_i are site energies and t_{ij} the hopping matrix elements generally assumed to be nonzero only between nearest-neighbor atomic sites $\langle i,j \rangle$. For an isolated molecule the basis is finite: both the Hamiltonian operator and the associated Green function are represented as matrices. Given the TB Hamiltonian for the molecule and a suitable model for the leads, it is thus possible to estimate the quantum conductance and, to a first approximation, the current–voltage characteristics of the device. The convenient framework provided by TB models for the computation of transport properties will definitely raise more and more interest in the description of transport through single polymers and in particular DNA wires using fast DFT algorithms based on localized orbitals. In this context, it is in fact possible to generate matrix elements that indeed represent a parameter-free version of TB Hamiltonians [129].

3.3.2
Applications to Poly(dG)-Poly(dC) Devices

The implementation of the scattering approach and of some simplified electronic structure models for describing the transport behavior of short poly(dG)-poly(dC) DNA wires [14] have been recently independently proposed within two main classes of models. One involves dephasing [123–125] and the other involves the hybridization of the π-stack [122].

The Porath et al. experiment [14], reviewed in Sect. 2, reports nonlinear transport measurements on 10.4-nm-long poly(dG)-poly(dC) DNA, corresponding to 30 consecutive GC base pairs, suspended between platinum leads (GC-device). DFT calculations indicated that the poly(dG)-poly(dC) DNA molecule has typical electronic features of a periodic chain [58]. Thus, in both models (assuming dephasing or π-stack hybridization) the poly(dG)-poly(dC) DNA molecule is grained into a spinless linear TB chain. A generalization of the dephasing model to spin-transport has been proposed by Zwolak et al. [123].

In the following we present the key components of these two models and show what consequences their assumptions have on transport.

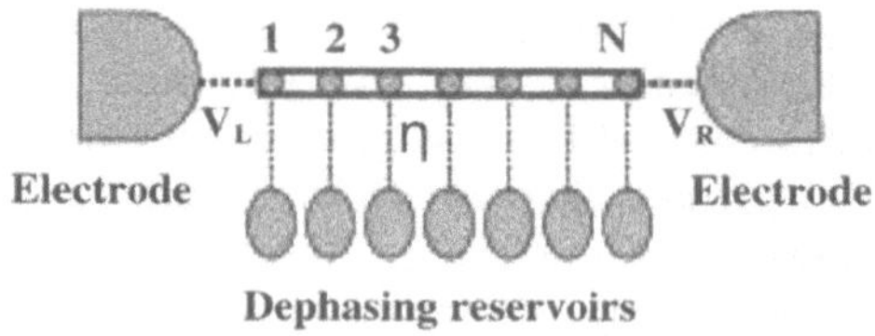

Fig. 21 The DNA molecule described by a one-dimensional TB Hamiltonian consists of a stack of GC pairs (the *circles*). To simulate the phase-breaking effect, each GC pair is connected with a dephasing reservoir (the *oval*) with a Hamiltonian $H_{\mathrm{deph}} = \sum_i \Sigma_i b_i^\dagger b_i$.. Following Ref. [132], all dephasing self-energies $\sum_i$ have a nonsmooth dependence on energy: $\sum_i = \eta^2/[E-\varepsilon_i-\sigma_i(E)]$, where $\sigma_i=(E-\varepsilon_i)/2-i[t^2-(E-\varepsilon_i)^2/2]^{1/2}$; the quantities ε_i and t are the linear chain TB parameters, and η is a parameter controlling the dephasing intensity. The total Hamiltonian thus reads as $H=H_{\mathrm{mol}}+H_{\mathrm{deph}}+H_{\mathrm{cont}}$ where the contact Hamiltonian H_{cont} accounts for the leads and their coupling to the molecule in a standard way [106] (from [125] with permission; Copyright 2002 by the American Institute of Physics)

3.3.2.1
Dephasing

The idea behind dephasing is that in complex structures, such as DNA, thermal motion and solvation effects might break the phase coherence of the system (a key assumption of the Landauer approach). Within Büttiker's treatment of inelastic scattering [130, 131], a possible way to account for dephasing effects in the Landauer approach consists of "inserting" dephasing reservoirs along the scatterer. This has been implemented by Li and Yan [124] by coupling a dephasing term to every site of the spinless linear chain describing the DNA wire. Figure 21 illustrates schematically the diagram for the model system.

3.3.2.2
Hybridization of the π-Stack

In a parallel development, Cuniberti et al. [122] suggested that a plausible mechanism for the observed gap in the Porath et al. experiment [14] could be the hybridization of the overlapping π orbitals in the G base stack with a perturbation source able to disrupt the metal-like channels. In this case the total Hamiltonian contains one term representing the spinless linear chain (the strong π-stack providing the coupling between any neighboring G-G pairs, as in Fig. 22), one term accounting for the hybridization of the π-stack (corrections to the π-stack strong coupling), and one term describing the contacts (Fig. 22). The hybridization term, which conserves the phase coherence, physically represents any interaction, internal or external to the molecule itself, which acts to break the perfect structure of the π-stack. Examples of such interactions are: (i) the transversal electronic degrees of freedom at each G base, possibly including the sugar group (Fig. 22); (ii) twist effects; and (iii) the inherent structure of the double helix which, with electrostatic

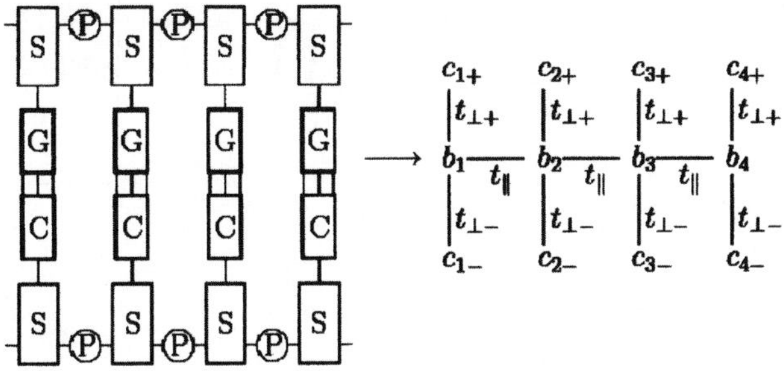

Fig. 22 Schematic view (*left*) of a fragment of poly(dG)-poly(dC) DNA molecule; each GC base pair is attached to sugar and phosphate groups forming the molecule backbone. On the *right side*, the diagram of the lattice adopted in building our model, with the π-stack connected to the isolated states denoted as $\pm$-edges. The total Hamiltonian reads $H=\sum_i \varepsilon_i b_i^\dagger b_i - \sum_{\langle i,j\rangle} t_{\|} b_i^\dagger b_j + \sum_{i.\alpha} \varepsilon_\alpha c_{i\alpha}^\dagger c_{i\alpha} - t_\perp \sum_{i,\alpha=} \left(c_{i\alpha}^\dagger b_{i\sigma} + h.c.\right) + H_{\mathrm{cont}}$, where the first two terms represent the spinless linear chain (the strong π-stack providing the coupling between any neighboring G-G pairs) and the second two account for the hybridization of the π-stack with the disconnected bands represented by the operator $c_{i\alpha}$. H_{cont} is the metal-molecule coupling Hamiltonian (from [122] with permission; Coypright 2002 by the American Physical Society)

and Van der Waals interactions, goes beyond a perfect "wire-like" channel where the coupling between neighboring planes may be defined purely within a linear-chain TB description. In the schematic view in Fig. 22, all these terms are gathered into the lateral (with respect to the transport main axis) coupling of the bases with two classes of sites $\alpha=\pm$, but the reader should bear in mind that the microscopic origin of this coupling is not specified at this level and may contain several distinct physicochemical effects.

3.3.2.3
Consequences on Transport

The dephasing and the π-stack hybridization are two different mechanisms responsible for the opening of a gap in the transmission. The latter assumes the particularly simple form $T=4\Delta_{\mathrm{L}}\Delta_{\mathrm{R}}|G_{\mathrm{LR}}|^2$ due to the low complexity of the underlying Hamiltonians [82]. Here, $\Delta_\nu=-Im(\sum_\nu)$ is the spectral density of the metal-molecule coupling at the left (L) and right (R) leads. G_{LR} is the molecular matrix element of the Green function between the two contact sites dressed by the lead self-energy[2].

[2] Interestingly enough the transmission calculation can be pursued analytically. In the particular case of the π-stack hybridized model [121], the gap in the transmission is a function only of the longitudinal and transverse hopping, $\Delta_T = 2\sqrt{t_{\|}^2 + 2t_\perp^2} - 2t_{\|}$. Note that in the case of nonhybridized π-stack $t_\perp$=0 the gap is closed, as we could expect. The gap opening mechanism (common feature of the two models) can also be understood by referring to their dispersion relations also with a gap at the Fermi level.

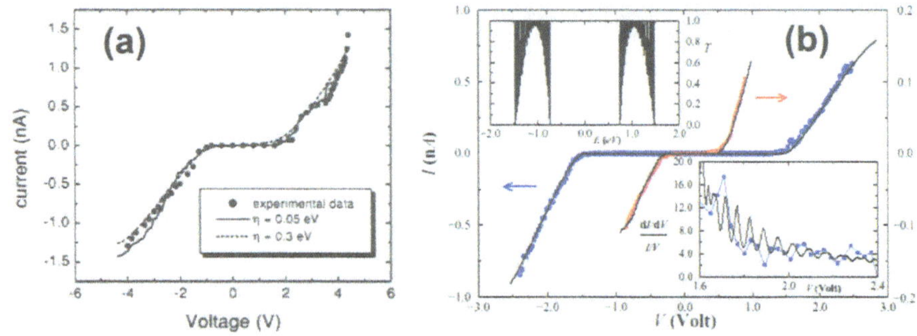

Fig. 23 **a** *I–V* characteristics from the work by Lee and Yang [124]: theoretical results versus experimental data. The transport currents in the presence of weak dephasing (η=0.05 eV) and a stronger one (η=0.3 eV) are theoretically shown by the *solid* and *dashed* curves. **b** Low-temperature *I–V* characteristics of two typical measurements for a 30-base-pair poly(dG)-poly(dC) molecule at 18 K (*blue circles*) and at 3.6 K (*red circles*) [122]. *Solid lines* show the theory curves following the experimental data. The *insets* show the transmission calculated after the blue data (*upper*) and the normalized differential conductance (*lower*). The parameters used (see Fig. 21) are $t_{||}$=0.37 eV and $t_{\perp}$=0.74 eV for the *blue* measurement, and $t_{||}$=0.15 eV and $t_{\perp}$=0.24 for the *red* one. These values are considered for a homogeneous system. If one draws the transversal coupling parameters for a disordered system their average value would be lower. For a related experiment supporting this idea cf. [113] (from [124] by permission; 2001 by the American Institute of Physics; from [122], with permission; Copyright 2002 by the American Physical Society)

Starting from the knowledge of the transmission $T(E)$, the calculation of the current can be pursued within the scattering formalism as presented in the previous section, and the results for the two models show an overall agreement with the experimental findings (see Fig. 23).

Despite the different assumptions in the two schemes described here, the properties of the calculated currents share the expected semiconducting features. This is basically due to the fact that they are formally acting on the π-stack in a similar way. Only further experimental investigation could shed more light on the discrimination of the gap-opening mechanism. Such an effort would eventually help in quantifying the influence of other insulating effects, which are possibly present in other DNA-based devices. They include (i) electron correlations and Coulomb blockade, (ii) localization effects due to the sequence variability, and (iii) local defects.

Although based on a simplified parametric description of the electronic structure of the molecule and of the leads, the framework discussed in this section has the advantage of leading directly to the computation of measurable quantities (the *I–V* curves). Thus, it is possible to relate the experimental observations to the quantum-mechanical properties of the systems under investigation, e.g., the electronic energy-level structure of the molecule and the relation of such levels to the energy of the leads. A timely improvement in this direction will come from the implementation of manageable methods, which combine a parameter-free atomistic description of the electronic structure of the molecular device [84–87, 129] with the well-established Green function formalism, to compute the transport characteristics. In such a way, it would be possible to distinguish self-consistently in the calculations different mechanisms, and their relative importance in experimental settings.

4 Conclusions and Perspectives

Charge migration along DNA molecules has attracted a considerable scientific interest for over half a century. The results of solution chemistry experiments on large numbers of short DNA molecules indicated high charge-transfer rates between a donor and an acceptor located at distant molecular sites. This, together with the extraordinary molecular recognition properties of the double helix and the hope of realizing the bottom-up assembly of molecular electronic devices and circuits using DNA molecules, has triggered a series of direct electrical transport measurements through DNA. In this chapter we provided a comprehensive review of these measurements and of the intense theoretical effort, in parallel and following the experiments, to explain the results and predict the electronic properties of the molecules, using a variety of theoretical methods and computational techniques.

After the appearance of some initial controversial reports on the conductance of DNA devices, recent results seem to indicate that native DNA does not possess the electronic features desirable for a good molecular electronic building block, although it can still serve as a template for other conducting materials [12, 13, 132]. Particularly, it is found that short DNA molecules are capable of transporting charge carriers, and so are bundles and DNA-based networks. However, electrical transport through long single DNA molecules that are attached to surfaces is blocked, possibly due to this attachment to the surface.

In a molecular electronics perspective, however, the fascinating program of further using the "smart" self-assembly capabilities of DNA to realize complex electronic architectures at the molecular scale remains open for further investigation. To enhance the conductive properties of DNA-based devices there have already been interesting proposals for structural manipulation. These include intrinsic doping by metal ions incorporated into the double helix (such as M-DNA [71–74]), other ions [119–120] exotic structures like G4-DNA [93, 118, 133], along with the synthesis of other novel helical structures. The reach zoology of DNA derivatives with more promising electrical performance will definitely represent a great challenge in the agenda of researchers involved with the transport in DNA wires.

A last comment is due on the need for progress in the investigation tools. Further development in the study of potential DNA nanowires requires the advancement of synthesis procedures for the structural modifications, and an extensive effort in X-ray and NMR characterization. Concerning direct conductivity measurements, techniques for deposition of the molecules onto inorganic substrates and between electrodes must be optimized. Moreover, the experimental settings and contacts must be controlled to a high degree of accuracy in order to attain an uncontroversial interpretation and high reproducibility of the data. On the theoretical side, a significant breakthrough might be the combination of mesoscopic theories for the study of quantum conductance and first-principle electronic structure calculations, suitable for applications to the complex molecules and device configurations of interest. Given this background, we believe that there is still plenty of room to shed light onto the appealing issue of charge mobility in DNA, for both the scientific interest in conduction through one-dimensional polymers and the nanotechnological applications. The high interdisciplinary content of such a research manifesto will necessarily imply a crossing of the traditional borders separating solid-state physics, chemistry, and biological physics.

Acknowledgements Funding by the EU through grant FET-IST-2001-38951 is acknowledged. DP is thankful to Cees Dekker and his group, with whom experiment [14] was done, to Joshua Jortner, Avraham Nitzan, Julio Gomez-Herrrero, Christian Schönenberger, and Hezy Cohen for fruitful discussions about the conductivity in DNA and critical reading of the manuscript. DP research is funded by: The First foundation, The Israel Science Foundation, The German-Israel Foundation, and Hebrew University Grants. GC acknowledges the collaboration with Luis Craco with whom part of the work reviewed was done. The critical reading of Miriam del Valle, Rafael Gutierrez, and Juyeon Yi is also gratefully acknowledged. GC research has been funded by the Volkswagen Foundation. RDF is extremely grateful to Arrigo Calzolari, Anna Garbesi, and Elisa Molinari for fruitful collaborations and discussions on topics related to this chapter, and for a critical reading of the manuscript. RDF research is funded by INFM through PRA-SINPROT, and through the Parallel Computing Committee for allocation of computing time at CINECA, and by MIUR through FIRB-NOMADE.

References

1. Luryi S, Xu J, Zaslavsky A (1999) (eds) Future trends in microelectronics: the road ahead. Wiley, New York
2. Joachim C, Gimzewski JK, Aviram A (2000) Nature 408:541
3. Aviram A, Ratner MA (1998) (eds) Molecular electronics science and technology. Annals of the New York Academy of Sciences, vol 852. The New York Academy of Sciences, New York
4. Aviram A, Ratner MA, Mujica V (2002) (eds) Molecular electronics II. Annals of the New York Academy of Sciences, vol 960. The New York Academy of Sciences, New York
5. Tour JM (2000) Acc Chem Res 33:791
6. Aviram A, Ratner MA (1974) Chem Phys Lett 29:277
7. Metzger RM (1999) Acc Chem Res 9:2027
8. Collier CP, Wong EW, Bolohradsky M, Raymo FM, Stoddart JF, Kuekes PJ, Williams RS, Heath JR (1999) Science 285:391
9. Reed MA, Zhou C, Muller CJ, Burgin TP, Tour JM (1997) Science 278:252
10. Collier CP, Mattersteig G, Wong EW, Luo Y, Beverly K, Sampario J, Raymo FM, Stoddart JF, Heath JR (2001) Science 289:1172
11. Chen J, Reed MA (2002) Chem Phys 281:127
12. Braun E, Eichen Y, Sivan U, Ben-Yoseph G (1998) Nature 391:775
13. Keren K, Krueger M, Gilad R, Ben-Yoseph G, Sivan U, Braun E (2002) Science 297:72
14. Porath D, Bezryadin A, de Vries S, Dekker C (2000) Nature 403:635
15. Rinaldi R, Biasco A, Maruccio G, Cingolani R, Alliata D, Andolfi L, Facci P, De Rienzo F, Di Felice R, Molinari E (2002) Adv Mater 14:1453

16. Rinaldi R, Biasco A, Maruccio G, Arima V, Visconti P, Cingolani R, Facci P, De Rienzo F, Di Felice R, Molinari E, Verbeet MP, Canters GW (2003) Appl Phys Lett 82:472
17. Rinaldi R, Branca E, Cingolani R, Di Felice R, Calzolari A, Molinari E, Masiero S, Spada G, Gottarelli G, Garbesi A (2002) Ann N Y Acad Sci 960:184
18. Alberti P, Mergny J-L (2003) Proc Natl Acad Sci USA 100:1569
19. Benenson Y, Paz-Elizur T, Adar R, Keinan E, Livneh Z, Shapiro E (2001) Nature 414:430
20. Benenson Y, Adar R, Paz-Elizur T, Livneh Z, Shapiro E (2003) Proc Natl Acad Sci USA 100:2191
21. Porath D, Millo O (1997) J Appl Phys 81:2241
22. Lemay SG, Janssen JW, van den Hout M, Mooji M, Bronikowski MJ, Willis PA, Smalley RE, Kouwenhoven LP, Dekker C (2001) Nature 412:617
23. Liang W, Shores MP, Bockrath M, Long JR, Park H (2002) Nature 417:725
24. Thorwart M, Grifoni M, Cuniberti G, Postma HWC, Dekker C (2002) Phys Rev Lett 89:196402
25. Lehn JM (1990) Angew Chem Int Ed 29:1304
26. Di Mauro E, Hollenberg CP (1993) Adv Mat 5:384
27. Niemeyer CM (1997) Angew Chem Int Ed 36:585
28. Niemeyer CM (2001) Angew Chem Int Ed 40:4128
29. Chen J, Seeman NC (1991) Nature 350:631
30. Zhang Y, Seeman NC (1994) J Am Chem Soc 116:1661
31. La Bean T, Yan H, Kopatsch J, Liu F, Winfree E, Reif JH, Seeman NC (2000) J Am Chem Soc 122:1848
32. Seeman NC (2001) Nano Lett 1:22
33. Zhang Y, Austin RH, Kraeft J, Cox EC, Ong NP (2002) Phys Rev Lett 89:198102
34. Dekker C, Ratner MA (2001) Phys World 14:29
35. Eley DD, Spivey DI (1962) Trans Faraday Soc 12:245
36. Warman JM, de Haas MP, Rupprecht A (1996) Chem Phys Lett 249:319
37. O'Neill P, Fielden EM (1993) Adv Radiat Biol 17:53
38. Retèl J, Hoebee B, Braun JEF, Lutgerink JT, van den Akker E, Wanamarta AH, Joenjie H, Lafleur MVM (1993) Mutation Res 299:165
39. Turro NJ, Barton JK (1998) J Biol Inorg Chem 3:201
40. Lewis FD, Wu T, Liu X, Letsinger RL, Greenfield SR, Miller SE, Wasielewski MR (2000) J Am Chem Soc 122:2889
41. Murphy CJ, Arkin MA, Jenkins Y, Ghatlia ND, Bossman S, Turro NJ, Barton JK (1993) Science 262:1025
42. Hall DB, Holmlin RE, Barton JK (1996) Nature 382:731
43. Kelley SO, Jackson NM, Hill MG, Barton JK (1999) Angew Chem Int Ed 38:941
44. Grinstaff MW (1999) Angew Chem Int Ed 38:3629
45. Barbara PF, Olson EJC (1999) Adv Chem Phys 107:647
46. Bixon M, Giese B, Wessely S, Langenbacher T, Michel-Beyerle ME, Jortner J (1999) Proc Natl Acad Sci USA 96:11713
47. Schuster GB (2000) Acc Chem Res 33:253
48. Conwell EM, Rakhmanova SV (2000) Proc Natl Acad Sci USA 97:4556
49. Okahata Y, Kobayashi T, Tanaka K, Shimomura M (1998) J Am Chem Soc 120:6165
50. Fink HW, Schönenberger C (1999) Nature 398:407
51. Nitzan A (2001) J Phys Chem A 105:2677
52. Meggers E, Michel-Beyerle ME, Giese B (1998) J Am Chem Soc 120:12950
53. Giese B, Amaudrut J, Köhler AK, Spormann M, Wessely S (2001) Nature 412:318
54. Davis WB, Naydenova I, Haselbeger R, Ogrodnik A, Giese B, Michel-Beyerle ME (2000) Angew Chem Int Ed 39:3649
55. Giese B (2002) Curr Opin Chem Biol 6:612
56. O'Neill MA, Barton JK (2002) Proc Natl Acad Sci USA 99:16543
57. Henderson PT, Jones D, Hampikian G, Kan Y, Schuster G (1999) Proc Natl Acad Sci USA 96:8353

58. de Pablo PJ, Moreno-Herrero F, Colchero J, Gómez Herrero J, Herrero P, Baró AM, Ordejón P, Soler JM, Artacho E (2000) Phys Rev Lett 85:4992
59. Storm AJ, van Noort J, de Vries S, Dekker C (2001) Appl Phys Lett 79:3881
60. Kasumov AY, Kociak M, Guéron S, Reulet B, Volkov VT, Klinov DV, Bouchiat H (2001) Science 291:280
61. Watanabe H, Manabe C, Shigematsu T, Shimotani K, Shimizu M (2001) Appl Phys Lett 79:2462
62. Shigematsu T, Shimotani K, Manabe C, Watanabe H, Shimizu M (2003) J Chem Phys 118:4245
63. Bezryadin A, Dekker C (1997) J Vac Sci Technol 15:793
64. Bezryadin A, Dekker C, Schmid G (1997) Appl Phys Lett 71:1273
65. Bockrath M, Markovic N, Shepard A, Tinkham M, Gurevich L, Kouwenhoven LP, Wu MW, Sohn LL (2002) Nano Lett 2:187
66. Gómez-Navarro C, Moreno-Herrero F, de Pablo PJ, Colchero J, Gómez-Herrero J, Baró AM (2002) Proc Natl Acad Sci USA 99:8484
67. Cai L, Tabata H, Kawai T (2000) Appl Phys Lett 77:3105
68. Rinaldi R, Branca E, Cingolani R, Masiero S, Spada GP, Gottarelli G (2001) Appl Phys Lett 78:3541
69. Lee HY, Tanaka H, Otsuka Y, Yoo K-H, Lee J-O, Kawai T (2002) Appl Phys Lett 80:1670
70. Tabata H, Cai LT, Gu J-H, Tanaka S, Otsuka Y, Sacho Y, Taniguchi M, Kawai T (2003) Synth Met 133:469
71. Rakitin A, Aich P, Papadopoulos C, Kobzar Y, Vedeneev AS, Lee JS, Xu JM (2001) Phys Rev Lett 86:3670
72. Aich P, Labiuk SL, Tari LW, Delbaere LJT, Roesler WJ, Falk KJ, Steer RP, Lee JS (1999) J Mol Biol 294:477
73. Wettig SD, Wood DO, Lee JS (2003) J Inorg Biochem 94:94
74. Li C-Z, Long Y-T, Kraatz H-B, Lee JS (2003) J Phys Chem B 107:2291
75. Yoo K-H, Ha DH, Lee J-O, Park JW, Kim J, Kim JJ, Lee H-Y, Kawai T, Choi HY (2001) Phys Rev Lett 87:198102
76. Jortner J (1976) J Chem Phys 64:4860
77. Marcus R, Sutin N (1985) Biochem Biophys Acta 811:265
78. Bixon M, Jortner J (2001) J Am Chem Soc 123:12556
79. Jortner J, Bixon M, Langenbacher T, Michel-Beyerle ME (1998) Proc Natl Acad Sci USA 95:12759
80. Bixon M, Jortner J (2000) J Phys Chem B 104:3906
81. Landauer R (1957) IBM J Res Develop 1:223, reprinted (1996) J Math Phys 37:5259
82. Datta S (1995) Electronic transport in mesoscopic systems. Cambridge University Press, Cambridge
83. Ferry DK, Goodnick SM (1999) Transport in nanostructures. Cambridge University Press, Cambridge
84. Xue Y, Datta S, Ratner MA (2002) Chem Phys 281:151
85. Calzolari A, Souza I, Marzari N, Buongiorno Nardelli M (2003) (preprint)
86. Brandbyge M, Mozos J-L, Ordejón P, Taylor J, Sokbro K (2002) Phys Rev B 65:165401
87. Gutierrez R, Fagas G, Cuniberti G, Grossmann F, Schmidt R, Richter K (2002) Phys Rev B 65:113410
88. Šponer J, Leszczynski J, Hobza P (1996) J Phys Chem 100:1965
89. Šponer J, Leszczynski J, Hobza P (1996) J Phys Chem 100:5590
90. Di Felice R, Calzolari A, Molinari E, Garbesi A (2002) Phys Rev B 65:045104
91. Calzolari A, Di Felice R, Molinari E, Garbesi A (2002) Physica E 13:1236
92. Adessi C, Walch S, Anantram MP (2003) Phys Rev B 67:081405(R)
93. Calzolari A, Di Felice R, Molinari E, Garbesi A (2002) Appl Phys Lett 80:3331
94. Gervasio FR, Carloni P, Parrinello M (2002) Phys Rev Lett 89:108102
95. Barnett RN, Cleveland CL, Joy A, Landmann U, Schuster GB (2001), Science 294:567
96. Dreizler RM, Gross EKU (1990) Density functional theory. An approach to the quantum many-body problem. Springer, Berlin Heidelberg New York

97. Hohenberg P, Kohn W (1964) Phys Rev 136:B864
98. Kohn W, Sham LJ (1965) Phys Rev 140:A1133
99. Becke ADA (1988) Phys Rev A 38:3098
100. Perdew JP, Chvary JA, Vosko SH, Jackson KA, Pederson MR, Singh DJ, Fiolhais C (1992) Phys Rev B 46:6671
101. Lee CT, Yang WT, Parr RG (1988) Phys Rev B 37:785
102. Becke ADA (1993) J Chem Phys 98:1372
103. Becke ADA (1993) J Chem Phys 98:5648
104. Johnson BG, Gill PMW, Pople JA (1993) J Chem Phys 98:5612
105. Cuniberti G, Großmann F, Gutiérrez R (2002) Adv Solid State Phys 42:133
106. Saito R, Dresselhaus MS, Dresselhaus G (1998) Physical properties of carbon nanotubes. World Scientific, Singapore
107. Elstner M, Hobza P, Frauenheim T, Suhai S, Kaxiras E (2001) J Chem Phys 114:5149
108. Voityuk AA, Jortner J, Bixon M, Rösch N (2000) Chem Phys Lett 324:430
109. Voityuk AA, Rösch N, Bixon M, Jortner J (2000) J Phys Chem B 104:9740
110. Voityuk AA, Jortner J, Bixon M, Rösch N (2002) J Chem Phys 114:5614
111. Bogár F, Ladik J (1998) Chem Phys 237:273
112. Ye Y, Chen RS, Martinez A, Otto P, Ladik J (1999) Solid State Commun 112:139
113. Friesner RA, Dunietz BD (2001) Acc Chem Res 34:351
114. Artacho E, Machado M, Sánchez-Portal D, Ordejón P, Soler JM (2002) airXiV:cond-mat/0209563 (to be published in Mol Phys)
115. Hjort M, Stafström S (2001) Phys Rev Lett 87:228101
116. Gottarelli G, Masiero S, Mezzina E, Spada GP, Mariani P, Recanatini M (1998) Helv Chim Acta 81:2078
117. Artacho E, Sánchez-Portal D, Ordejón P, García A, Soler JM (1999) Phys Status Solidi B 215:809
118. Phillips K, Dauter Z, Morchie AIH, Lilley DMJ, Luisi B (1997) J Mol Biol 273:171
119. Tanaka K, Yamada Y, Shionoya M (2002) J Am Chem Soc 124:8802
120. Tanaka K, Tengeiji A, Kato T, Toyama N, Shionoya M (2003) Science 299:1212
121. Carloni P, Andreoni W (1996) J Phys Chem 100:17797
122. Cuniberti G, Craco L, Porath D, Dekker C (2002) Phys Rev B 65:241314(R)
123. Zwolak M, Di Ventra M (2002) Appl Phys Lett 81:925
124. Li XQ, Yan Y (2001) Appl Phys Lett 79:2190
125. Li XQ, Yan Y (2001) J Chem Phys 115:4169
126. Imry Y (2002) Introduction to mesoscopic physics, 2nd edn. Oxford University Press, Oxford
127. Fisher DS, Lee PA (1981) Phys Rev B 23:R6851
128. Jauho AP (2003) Nonequilibrium Green function modeling of transport in mesoscopic systems. In: Bonitz M, Semkat D (eds) Progress in nonequilibrium Green's functions II. World Scientific, Singapore
129. Frauenheim T, Seifert G, Elstner M, Niehaus T, Köhler C, Amkreutz M, Sternberg M, Hajnal Z, Di Carlo A, Suhai S (2002) J Phys Condens Matter 14:3015
130. Büttiker M (1986) Phys Rev B 33:3020
131. D'Amato JL, Pastawski HM (1990) Phys Rev B 41:7411
132. Richter J, Mertig M, Pompe W, Mönch I, Schackert HK (2001) Appl Phys Lett 78:536
133. Parkinson GN, Lee MPH, Neidle S (2002) Nature 417:876

Author Index Volumes 201–237

Author Index Vols. 26–50 see Vol. 50
Author Index Vols. 51–100 see Vol. 100
Author Index Vols. 101–150 see Vol. 150
Author Index Vols. 151–200 see Vol. 200

The volume numbers are printed in italics

Subject Index

MIX
Papier aus verantwortungsvollen Quellen
Paper from responsible sources
FSC® C105338

If you have any concerns about our products,
you can contact us on
ProductSafety@springernature.com

In case Publisher is established outside the EU,
the EU authorized representative is:
Springer Nature Customer Service Center GmbH
Europaplatz 3, 69115 Heidelberg, Germany

Printed by Libri Plureos GmbH
in Hamburg, Germany